BROKEN NATION

VOL 2

BOUND BY WEAK TIES

BABA OLU

Brokenation

Brokenation, Voliume 2 - Bound by Weak Ties.
www.thebrokenation,com

Printed in the United States of America

THE PURCHASE OF 'NIGERIA'

From the West Australian, July 5, 1899

ROYAL NIGER COMPANY - IMPERIAL PURCHASE OF NIGERIA. LONDON, July 4;

The House of Commons yesterday by 223 votes to 101 agreed to the Vote of £865,000, with which it is proposed to purchase the rights of the Royal Niger Company in "West Africa".

Sir Michael Hicks-Beach, while praising the company's administration of the territories embraced in their charter, admitted that the Anglo-French relations in regard to West Africa since 1897 had been more strained than the public had appreciated.

Owing to the Royal Niger Company having had a free hand, a terrible war had been threatened. The Niger territories, now called Nigeria, have been administered by the Royal Niger Company under a Royal charter dated July 10, 1836. They cover about 500,000 square miles and contain a population variously estimated at from 20,000,000 to 40,000,000. In October last year it was reported that the Imperial Government had taken over Akassa from the company, and the terms upon which the territories were to be transferred to the Imperial Government have been made the subject of many rumours.

From the Auckland Star, July 5, 1899

ROYAL NIGER COMPANY - IMPERIAL PURCHASE OF NIGERIA. LONDON, July 4:

The wisdom and justice of allowing-companies established mainly for money-making purposes to make war upon native tribes and establish governments of an autocratic character over vast possessions of the Crown may well be questioned. The methods of some of these companies are not characterised by fine scruples, and whenever their interests conflict with those of the native, tribes inhabiting the lands under their rule the black man receives scant justice.

So long as the wars arising out of the operations of these trading companies are confined to expeditions against bloodthirsty negro potentates the Home Government are not too inquisitive regarding the methods of warfare adopted, and the news received from time to time that thousands of blacks at some place in the African wilds have been mown down by Maxim guns served by a handful of British troops is usually hailed with popular acclamation as another evidence of our national prowess. But when the action of these chartered companies brings the empire into conflict with other European States which are. pursuing the same game then the matter assumes an entirely different complexion.

For several years past the Royal Niger Company has been beset by French encroachments on the North and West, and German operations on the East. The differences with Germany were easily settled owing- to the reasonable and amicable disposition manifested by the German Government, but France, in Nigeria, as in Egypt, has manifested a disposition to flout "perfidious Albion." Three or four agreements on the boundary question have been' made, but each one, so far as French claims went, left certain boundary lines in an indefinite state, and in a country over which the administration is of the loosest description, many parts being rarely' visited even by trading agents, it is easy for explorers and adventurers to enter, and on the. strength of treaties with native kings who have not the smallest notion what they are signing to establish a basis for national claims. This has gone on for many years in Nigeria, to the great disturbance of friendly relations between England and France; the announcement, therefore, made some time ago that all outstanding' points of difference had been amicably settled was a relief.

To prevent future complications and also to strengthen Great Britain's hold over the Niger territories the Imperial Government have now resolved to purchase out the rights of the Chartered Company for £560,000. It seems a big price, but is after all a trifle compared with the cost of a European war, which the Chancellor of the Exchequer declared had been more nearly threatened than the public appreciated through the free hand exercised by the company.

PREFACE

Frederick Lugard, a dismissed English Army Officer seeking redemption and fame, had proved useful to Colonial ambitions as sanctioned, but unofficial armed muscle in Karonga (Malawi) and Buganda (Uganda) in East Africa, before his invitation to West Africa by the Royal Niger Company, where his style and ambition once again aligned with English colonial interests.

The West African frontier force, which at times was under his direct control, gave him the leverage needed for his own objectives, which were sometimes at odds with those of the more experienced colonial officials.

The (il)legal acquisition of Nigeria by the English Imperial Government from the Royal Niger Company was effectively a takeover of the English Colonial structure in "Nigeria" by Lugard.

This only provided gasoline to fuel Lugard's quest for fame in his homeland though, and despite the colonial offices recommending various independent regions to reflect the complexity of Nigeria (not that they had any moral or legal right to slice, dice and glue pre-existing societies and nations), Lugard pushed for the most simple administrative setup and forced various countries into two protectorates to give him enough time (roughly about half of the year) to vacation in England, speaking at parties and social events about his African adventure.

This was to facilitate his desired climb up the English social ladder, which had been derailed by his moral indiscretions.

ACKNOWLEDGEMENT

To God, the Father whose name is hallowed
The Almighty, All knowing Sovereign
Who subjects Himself to Righteousness and Justice
and as such to whom the means matter as much as the end

DEDICATION

To

Wing Commander Chimda Stephen Hedima
(October 1, 1977 - Circa October 1, 2014), captured as his aircraft came under fire, while providing cover to his at-risk and under fire Army comrades on the ground and his copilot, Group Captain Abdulrasheed Bamidele Braimoh, still reported as Missing in Action as of 2021.

To their comrades who have lost their lives, and to those maimed and injured in action and to the thousands of Officers, Men and Women of the Nigerian Army, Navy and Air Force, swimming against the tidal wave of faulty foundations. Not asking questions, but serving, protecting, and preventing the innocent and defenseless from being completely overrun.

Your predecessors chased the Italians out of Ethiopia and the horn of Africa, and the forgotten 81st and 82nd Divisions that stalled the Japanese in Burma.

CONTENTS

PART 3

A Short History of the Nigerian Military, and the Nation It Helped Forge

The history of the current Nigerian State can be viewed through the history of Nigeria's Military. That view is clear in most aspects and blurry in a few others, but it can be argued that the Nigerian Military has been the most significant force and factor, in the history, politics, and indeed the very existence of the Nigerian nation, since Flora Shaw coined the name 'Nigeria' in 1897, and in the few decades before that moment. Shaw's naming move helped distinguish the 'French owned' Upper Niger areas (Niger), from the 'English owned' Lower Niger areas, with the Hausa country split almost down the middle between the two 'European-owned' territories.

The impact of the Military on Nigerian existence has been so deep and wide, that a cursory reference on its sometimes unifying and/or defining role, will be grossly inefficient, as significantly more time and words will have to be expended to cogently capture and communicate this impact.

Men of the military have been used (and have sometimes acted on their own, or collaboratively) to chart and alter the course of the nation for good, and in equal or greater measure, for bad.

As such, what was meant to be a subsection of "The Weak Ties that bind" has transformed to be a mini-history book, which might not be a bad thing for the millions of Nigerians who passed through school with no relevant and contemporary History as part of the curriculum[1], in what can only be described as an attempt to suppress the truth by wickedness. This will however still be a summarized version at best, and this is an area begging for much more research, documentation, presentation to the public and for posterity.

The roots of the Nigerian Military lie in the Lagos Constabulary, also known informally as Glover's Hausas which was formed in 1864. This unit either morphed or was repackaged into new, related or successor units were set up over the next decades, to meet specific needs of the English Colonials (Government and Royal Niger Company), so names like the 'Royal Niger Company's Constabulary and the West African Frontier Force which emerged, can be put in this context.

What was clear about these forces was that while they were mostly made up of indigenes of the local colonies and protectorates (especially ethnic Hausas), these constabularies and frontier forces were set up for the interests of foreigners and non-indigenes, and were fully under their control. Some would justifiably regard today's Nigerian Army, still bearing the insignia of the invading Foreign Fulbe Fuduyan militia, as essentially the same in spirit and purpose.

1 https://guardian.ng/features/re-introduction-of-history-and-matters-arising/

LUGARD AND THE WEST AFRICAN FRONTIER FORCE

The events and factors that led to the creation of what eventually became Nigeria's military were rooted in the interests and actions of non-indigenous parties. Firstly, the Fuduyan takeover of Gobir, in Hausa Country, led to a swift herdsmen-aided takeover of the rulership of the Hausa country (city states split across the current borders of Nigeria and Niger Republic) by some Fulbe clans within a few years.

This led to a significant internal (Maguzawa) and external displacement of Hausas, especially men, many single, all over modern-day Nigeria and other parts of West Africa.

The takeover of the political, traditional and religious leadership of Hausaland also caused a fundamental shift in the socio-economic structure of Hausaland, which led to the constant outpouring of young men from their country, and this has largely continued unabated for 220 plus years. In the decades before the formation of the Lagos Constabulary, the Alaafin of Oyo's Calvary had become almost entirely made up of Hausa men due to this phenomena, and the colonials must have picked up on this and gone for the 'lowest hanging fruit' in the composition of their forces. It also helped because due to the long history of commerce of the pre-Fulbe Hausa states, Hausa had become a second language for many non-Hausas including many parts of the area referred to as the 'Middle Belt' or 'North-Central' in today's (2021) Nigeria.

A subtle, but fundamental impact of all this, was that you did not necessarily need to go to Hausa country to recruit young Hausa and Hausa speaking men.

Some of the displaced Hausas formed the bulk of the Lagos constabulary when they were formed in 1864. *Glover's Hausas* and saw action in Lagos and in the current western Nigeria and in Ghana, and were a significant part of what later became the West African Frontier Force, despite the usual politics and territory-claiming that existed between the various colonial governments who were afraid of the dominance of the 'Nigerian' contingent and section of the force.

The second major factor in which Nigeria's military history is rooted is the commercial interest of England's United Kingdom (endonymically referred to as Great Britain). No man represented this more than Frederick Lugard, an officer whose promising army career had been truncated for having an affair with a married woman. Seeking redemption, and a fresh start after this shameful event, he went to East Africa (Bugandaland) and took a very different approach (as semi-sanctioned armed muscle promoting English interests) to "continuing the work of David Livingstone" in East Africa.

Lugard had a get-famous-or-die-trying mentality, and the interest in the works of David Livingstone in British and Western society (or in the English-speaking world) from a few decades earlier proved to be a path of least resistance for Lugard to reverse his fall from grace and achieve the place in British high society he so desperately desired. Unlike Livingstone, Lugard had no missionary or medical interests in the service of the locals and his military background made him ideal as the kind of armed muscle to force things through in the pursuit of English commercial and colonial interests.

Lugard's membership of the freemason (the occult in Nigerian parlance), has also quite naturally brought spiritual and logical deductions about the role he was eventually to play in Nigeria's history. His success and reputation as an enforcer in East Africa, led to his employment by the Royal Niger Company, which was pursuing British commercial interests in what is now Nigeria, to reprise his role in Bugandaland. Specifically, he was invited to force the French to beat a retreat from Borgu.

Lugard saw the West African Frontier Force as an army over which he would have ultimate control and
had been quoted as saying that it would "be employed to serve beyond Her Majesty's dominion," which was taken to mean as a mercenary force subject to his own personal desire and control. He obviously had his own agenda, which obviously had large areas of intersection with English interests, but was often aside and beyond, and did not accommodate any interference even from the colonial office.

Lugard's Borgu expedition had mixed results, with a significant part of the Borgu kingdom 'lost' to France (in the current Republic of Benin). He however veered off-course and went on a personal mission to dislodge the Fulbe Clanleaders and rulers of Hausaland, interestingly, mostly with Hausa soldiers who had been displaced by the overthrow of the previous Sarkins of the city-states of the Hausa Country.

Quite remarkably, some cases of direct confrontation between Lugard's WAFF and the Fulbe rulers of Hausaland and Ilorin led to the flights of the rulers from these domains, which was in sharp contrast from the responses of the indigenous rulers like the Awujale in Ijebu, Jaja in Oopbo and the Oba of Benin who stayed put in their kingdoms, come what may, when in direct confrontation with the English forces.

Despite the claim of wanting to remove the slave raiding Fulbe 'Emirs' (as as the Clanleaders referred to themselves to put a religious or Arabic hook on things) of Hausaland, Lugard restored virtually all them back to the rulership of Hausaland, with the final act being the restoration of the leadership of the Fulbe Clans to Attahiru in Sokoto, after the pursuit of the similarly named Attahiru, the incumbent leader and brother of the new leader, who had been killed in Gombe by Lugard's pursuing forces.

For believers in matters beyond the temporal, there have been some questions about the spiritual connotations, if any, of Lugard's actions in Hausa Country as Uthman Fuduye, the leader of the Fulbe takeover of Hausaland had predicted the rulership of his clans would be displaced after a hundred years by another group of foreigners, and his Fulbe clans were to go set up a puritan religious state by the Nile, in the Sudan, when this happened.

The Fulbe leader Attahiru must have been conflicted by Lugard's curious decision to abandon his mission in Borgu and face the Fulbe Clanleaders of Hausa land. On one hand, he knew the time was up, based on his progenitor's directive; on the other hand though, it appears the benefits of sovereign rule were too good to give up (even as the subjugated country got progressively poorer and desolate) and he tried to have it both ways.

He eventually sent his son Mai (W)'Urno to fulfill the ancestral directive to migrate to the banks of the Nile in Sudan, as Lugard's forces closed in on the fleeing party in current day Gombe.

APES'S OBEY

Let's take a brief interlude from the impactful, and nation-enforcing history of Nigeria's military and project about 83 years into the future from the sacking of Attahiru in Burni in 1903 to a Military School in Nigeria where two boys (IK & BJ) seemed to argue all the time over all manner of things ranging from grammar and pronunciation (how is the McVities 'TUC' biscuit pronounced anyway?) to life and living. Unlike many of their other classmates though, they didn't really discuss and argue about the Nation, the military and coups. The 'barrack boys' (sons of Officers and Men of the Armed Forces) tended to do that more, given their background. It was nonetheless a life of complete immersion into the military training and mindset with endless drills, parades, weapons training, field training, route and compass marches, obstacle crossings and map reading (thankfully made obsolete by GPS) and endless runs and PT; and this was only the official part of it. We'll skip the unofficial, more-intense bits.

Quite naturally, a lot of the activity was encouraged by songs and chants, many of which most of the boys just memorized without knowing the meaning, especially since the songs were usually in Hausa, although there were a few chants from other parts of the Nigerian Nation (Got to have 'Enyimba Enyi' and 'Kerenke Obi' in any Nigerian Military Cadence collection).

There was a particular chant though, that made IK & BJ get into another verbal confrontation, but which had a deeper and fundamental meaning on the history of the Nigerian Military. To BJ's ears, it sounded like 'Ehtsobey', even though that meant nothing, but so did several other chants and songs in the military. So, when IK huddled the group of 5 one day and asked that all 5 stop responding to that particular chant (at the risk of getting into trouble, depending on the 'type of parade'), everyone waived him off as an 'ITK', even though no one from the group of 11 year olds could give an alternative meaning to counter his claim that it was a condescending colonial chant mocking the illiterate (in English) soldiers of the WAFF, by their English NCOs and Officers.

BJ thought IK might have a point, but could not get himself to accept that one group of people would mock, insult, or take advantage of another group of people, just for kicks, and right in their faces. There had to be an alternative explanation, there just had to be. Maybe it was a chant that had its origins in a minor Nigerian language, or even a foreign language (it had a South African liberation chant and vibe to it). So he asked all his 'non-WAZOBIA' classmates if 'Ehtsobey' meant anything in their language.

It didn't and it doesn't. Oftentimes (and some will argue, most of the time), the truth is the simplest and most obvious explanation, and it is staring at you, right in the face.

The fact that the indigenously-peopled Frontier Force had brutally, ruthlessly and efficiently subjugated the country did not earn it respect or even basic dignity from its foreign assemblers. Quite curiously, it is suggested that Lugard, by reinstalling the Fulbe clan leaders of Hausaland, handed the country right back to another group of foreigners. It was however not Lugard's country to give in the first place; However Fuduye's insignia still adorns the logo of the Nigerian Army in 2021.

That is the tragedy of colonialism, and allowing foreigners for whatever reason to rule over you. There will always be a reason or hook (religious, political, economic, civilization, development etc) used to justify the takeover or occupation, but at its core is a deliberate attempt to strip a people of their independence, dignity, history, culture, freedom, way of life and and to some extent part of their humanity.

POST AMALGAMATION

By 1914, the West African Frontier Force had grown to include 244 English officers, 115 English NCOs and 7,201 native officers. England's United Kingdom was now fully in control of 'Nigeria' and, the WAFF had to adjust from being an army of violent subjugation to pursuing and protecting English interests, somewhat following the path of the Fulbes clans before them who transitioned from violent propagations to industrial scale slave raiding and trading.

Even though the composition of the West African Frontier Force was indisputably mostly Hausa, there were some non-Hausa NCOs who had taken up leadership positions in this Force, perhaps due to their ability to speak the English language or their exposure to missionary education by the 1920s. Amongst these were the highly decorated RSM Belo Akure, Belo Ojo and Inspector Ajayi, who were part of the exhibition of the British Empire in Wembley. Some have alternatively suggested that RSM Akure & Ojo may have been ethnic Hausas resident in 'Western' Nigeria from the continuous displacement from the Fulbe takeover of Hausaland. This is plausible given the Belo first names, and the tendency of Hausa to be surnamed after cities in which they reside (Akure for instance).

RSM Belo Akure, Sgt Belo Ojo & Insp. Ajayi at the British Empire Exhibition at Wembley in 1924.

Over 45,000 Nigerian soldiers served in the Second World War, and the bravery of the Nigerian soldiers was renowned. English recruitment for the war effort was easy, at least at the beginning with the full support of colonial mechanism, and nationalist and independence leaders like Herbert Macaulay and Nnamdi Azikiwe. A significant number of these soldiers were involved in bloody battles against the Japanese in Burma and Sri Lanka, but also played significant roles on the African continent, especially in the capture of Mogadishu (the current capital of Somalia). Nigerians also literally chased the Italians out of Ethiopia. All of these roles are hardly spoken about neither by the colonials nor by citizens of the countries that were liberated by the Nigerian Soldiers.

Nigerian soldiers were not even acknowledged or mentioned in the victory speech by the commanding officer, William Slim at the end of the War. Their experience was however better than that of the Senegalese soldiers who, likewise, fought on behalf of the French, and upon return to Senegal, were held in prison and gunned down in Thiaroye over protests about pay and pensions.

It is noteworthy that some of the battalions of the Nigerian Defence Academy including Abyssinia, Burma, and Mogadishu are named in commemoration of some of the exploits of Nigerian soldiers during the Second World War. It is also noteworthy that while some of the previously single campus institutions like the Nigerian Law School now have campuses spread all over the country, the Nigerian Defense Academy remains a single campus institution in Kaduna.

The use of mostly Hausa rank and file soldiers to depose and later reinstate the Fulbe rulers of Hausa land was not lost on the ruling clans, and most especially on Ahmadu Rabbah (Bello), leader of the Northern Peoples Congress and defacto ruler of post-independence Nigeria, and would largely frame their opinions, thoughts, actions and approach to Nigeria's independence and the period immediately afterwards. Rabah Bello enjoyed a rapid rise in politics, leveraging on the Fulbe control of the traditional and religious institutions in Hausaland (Nigeria's largest indigenous country), and his background as a leading member of the ruling clans. He had lost out in the effort to become the head of the clans to his relative Siddiq Abubakar, but soon found that the political process in the emerging Nigerian state was a much more effective means to exert power and control.

After their service with distinction during the 2nd World war, Nigerian soldiers started to demand for recognition and inclusion in the Officer Corps of the military, especially after further distinguished service as part of the United Nations Peacekeeping Force in the Congo. Many of these soldiers, especially the southern ones, feeling a little ill-treated by the British, supported the independence efforts.

Queen Elizabeth of England reviewing a Guard of Honour during a visit to Nigeria.

However, the 'Other Ranks' (NCOs) of the military was still mostly dominated by Hausas, both from the north and those resident in the south, but with a growing role for the middle belt, especially with Tivs, Idomas, Birom and Tarok NCOs. The Officer Corps, due to the education and qualification standards required, was however dominated by Southerners, and this proved to be a huge problem for Ahmadu Bello.

Left to right sitting: Captain Robert Adeyinka Adebayo, Captain Philip Effiong, Captain Umeh Ogere Imo, Major Samuel Adesoji Ademulegun, Major Wellington Bassey, Major General Norman Forster (GOC, Nigerian Army), Major Aguiyi Ironsi, Major Ralph Adetunji Shodeinde, Captain Zakaria Maimalari, Captain Conrad Nwawo, Captain David Akpode Ejoor. 2nd Row Standing: Lt Igboba, Lt George Remunoiyowun Kurubo, (non Nigerian standing next to Kurubo), Lt J Akahan Akaga, Lt Patrick Awunah, Lt Louis Ogbonnia, Lt Chukwuemeka Odumegwu Ojukwu, Lt Eyo Ekpo, Lt Author Unegbe, Lt Abogo Largema. 3rd Row Standing: Lt Hillary Mbilitem Njoku, 2nd Lt Macauley Nzefili, 2nd Lt David Ogunewe, 2nd Lt Shadrack, Lt Alexander Madiebo, 2nd Lt Anthony Eze, Lt Yakubu Gowon, 2nd Lt Sylvanus Nwanjei, Lt Yakubu Pam, 2nd Lt Hassan Katsina.

POST INDEPENDENCE – QUOTAS OVER MERIT, THE MILITARY FOR POLICE DUTIES

Nigeria's post-Independence Prime Minister, Abubakar Tafawa Balewa insisted that the country would stick to the principle of merit in choosing the men for the officer corps of its military. This was at complete odds with his party leader Sir. Ahmadu Bello, who had declined to become Prime Minister, partly due to his disdain for Lagos, having failed to convince the British to move the capital to Kaduna (the feeling was mutual as Lagosians had insulted Bello as the man who kept them under British rule, when the rest of the country was ready to become independent). He promised to return to Lagos, "with a sword in his hand" when next he came in response to those insults, and Bello, and Defence Minister Muhammadu Ribadu (both of these gentlemen had an interesting past with claims of cattle acquisition) severally pressured the Prime Minister to abandon the principle of merit in the acceptance of men to the Officer Corps of the Military, but the Prime Minister, perhaps due to his sensitivities as an indigenous Nigerian, as opposed to a foreign Fulbe, stood firm and stuck to his position.

Balewa's Defence Minister, Ribadu helped manage the sometimes complex relationship between the Prime Minister and Ahmadu Bello. Some believe he would have found a way to avert the crisis of 1966, if he was alive. His Granddaughter, Aisha later married Muhammadu Buhari one of the boys he helped get into the military on the principle of 'stability'.

Upon the completion of the Nigerian Military (now Defence) Academy, Kaduna, Nigeria could now locally train its military officers instead of sending them to military colleges abroad. However Ahmadu Bello was livid that many young men that he and the 'Northern' ruling class encouraged to seek admission into the academy were refused on the basis of lack of proper academic or medical fitness. Bello was very concerned that only six of Nigeria's 44 Military officers at the time were Northerners, with some of these Northerners being Christian Benue-Plateau officers.

At this point, the Prime Minister gave in to pressure, despite his previous stance about such a policy being unfair and leading to an erosion of standards. Defence Minister Ribadu's statement that "stability was at least as important as fairness" could not have been comforting to the Prime Minister. As can be expected from actions based on such flawed thought process and logic, neither Nigeria nor its Military has had any enduring stability, equity (fairness) or standards since then. The Army thus saw an influx of Northern officers, based on quotas meant to help the region 'catch up', and not the regular qualification criteria for the officer corps. Several of these 'Bello-Ribadu' officers were to play leading roles in the coups, counter coups, and Military Governments (including as Heads of State) that plagued the country over the next five decades.

Ben Gbulie, Nigerian Army, and later Biafran Army Officer, gave specific examples of this impunity:

> *In an attempt to catch up militarily with the South, the Northern politicians had thrown out all discretion. They had lowered standards of admission drastically, settling for the minimum. For as I recalled, all the Northerners in my intake had been trained at the Mons Officer Cadet School in Aldershot. And they had become officers after barely six months of military training, whereas those of us who had been sent to Sandhurst had had to do two long years to earn the Queen's Commission. The implications were quite clear and most disturbing. Not only had these Northerners become commissioned officers before we were half-way through our first year at Sandhurst, they had all risen to the enviable rank of Captain before we could even appear at the sovereign's parade which served essentially*

as a prerequisite for our passing out as Second Lieutenants."

"By 1964 a group of young Nigerian officer-cadets, mostly Northerners, had been declared academically unfit and hence repatriated by the Canadian military authorities. These cadets were however pronounced commissioned by the Nigerian Federal Government no sooner than they had arrived at the Ikeja Airport. Consequently they had had to be absorbed into the Nigerian Army as commissioned officers, even though they had received no requisite military training."

A funny side note to the Ribadu/Bello policy was that a tiny sub-section of those who made it into the Army through this policy were not from the intended ethnic groups, but were able to mask their identities with assumed names, and the ability to speak Hausa, usually by virtue of being born or raised in Hausa Country.

Ribadu was extremely useful in managing the sometimes complicated relationship between the Prime Minister, Balewa and his party leader and Northern Premier, Bello, who increasingly seemed to discount and disregard National affairs and the office of the Prime Minister. He'd severally publicly state that he'd rather be Sultan of Sokoto (the lead office, with the strong religious reference, of the ruling Fulbe clans of Hausaland) than Prime Minister. Bello was also known to react angrily in being mistaken for being Hausa (especially in the South), and was known to angrily interrupt conversations that innocuously identified him as Hausa[2] with responses like "I am not Hausa. I'd rather be dead than be Hausa."[3]

Ribadu, as Defense Minister, oversaw the start of the Air Force and Navy, mostly through the transfer of service of regular Army Officers. A close camaraderie remains between the Officers of all three Armed Forces as almost all the 'regular' officers are admitted and trained together for at least four years at the Nigerian Defence Academy (before the Naval Cadets head out for a final year of Sea training as Midshipmen). He also helped ensure that Kaduna (and nearby Zaria) was the focus for Nigeria's military infrastructure and institutions, except for the Navy, for obvious reasons.

Ribadu's prior experience as Minister of Land, and of Lagos affairs, meant that he was a little bit more appreciative of the needs and sensitivities of the Federal cabinet, even if his primary consideration was to his party and the Fulbe Feudalist hegemony. He however had a strong influence on the NPC leader, and was able to get him to change his mind when necessary and make life easier for Balewa. Ribadu's death in 1965, complicated things for Balewa especially, as his replacement as the private go between both leaders, Inua Wada, did not seem to have the same influence and persuasive qualities over Bello, who seemed increasingly more focused on his interesting religious preoccupations, many in collaboration with Abubakar Gumi, religious scholar and critic of Siddiq Abubakar. Bello performed a very public (re) conversion of Dentist and Gambian Prime Minister, David (Dauda) Jawara.

The normally calm and conciliatory Prime Minister was publicly quoted as saying, "I'm tired, go and tell them to do

---

2 Rt. Hon Gentleman, Page 651

3 Honour in African History, Page 314

whatever they want," in frustration to another Bello move that had put him in a tight spot.

BEYOND THE MILITARY HISTORY

It will probably help to broaden our perspective of the emergence of the Nigerian State from our strict focus on history of its Military, and look at other critical factors, players, events, motives, decisions and institutions that helped chart the course of a Nation cobbled together for reasons that have nothing to do with the benefit of its indigenous peoples and constituent countries.

The world was a very different place at the end of the 2nd World War, and for an England that was probably saved from a German occupation by a combination of American involvement and the Soldiers from its colonies (almost 1.5 million from India alone), the Indian Independence Act of 1947 was a huge problem.

The fear and possibility of a 3rd World War was very real at this time, and the lack of access to 1.5 million troops that could be called from newly independent India meant the English Colonials needed an 'insurance policy' from where a million or two million troops could be quickly assembled and trained for its service if needed. The place that best fit this profile in the British Empire was Hausa Country under the Fuduyan Fulbe rule.

Most of the men in the Lagos Constabulary that transformed to 'Glover's Hausas' and the West African Frontier Force, were ethnic Hausas, as were a significant number of the 45,000 'Nigerians' that served in World War 2 (WW2). If the English Colonials could get this number of Hausas without a direct and aggressive recruitment in Hausaland, then they could definitely get a million or two if directly collaborating with the Fuduyan Fulbe rulers of Hausaland.

It was unlikely that any other part of Nigeria would offer her sons as canon fodder for the English cause, and Hausaland under the rule of indigenous Hausas would have been a big gamble. It was thus important that Hausaland remained under the rule of the foreign Fulbe clans. In this regard, Lugard's decision to re-install the Fulbe Clanleaders in Hausaland 40 plus years earlier proved fortuitous (it is unlikely that it was planned for this purpose, but it can't be ruled out as the war mongering European powers were on the verge of WW1 at this time).

Taking aggressive self-preservatory decisions was not uncommon for the English, during and in the period after the 2nd World War. For instance, Churchill (through his Secretary of State for Colonial Office, one of the sources quoted in the 'From the Horse's Mouth' section) offered to cede Northern Ireland to Ireland, if Ireland abandoned its stance of Neutrality and joined WW2 on the English side. Ireland however declined thinking that keeping most of Ireland Irish, was better than losing all of Ireland (and the United Kingdom) to Germany, which was common thought at the time, before the entrance of the United States into WW2 in December 1941.

This mindset of the English towards Nigeria, also led to the concentration of military training and the bulk of military activity to Kano, Zaria, and Kaduna City (though just outside or on the fringe of Hausa country).

FROM THE HORSE'S MOUTH

Official Transcripts and Records of the expressed thoughts, conversations, opinions and motives of High-level Colonial and English officials on the events and personalities of 'Pre-Independence' Nigeria.

They are sourced from the 'British End of the Empire Project', edited by Martin Lynn.

Official records in Italics, BN author's comments highlighted.

Evidence that the English Colonials set up Bello's invitation to the Action Group in 1952.

July 1958

[Political situation]: minutes by M G Smith, A Emanuel, Sir J. Macpherson, and Lord Perth on the implications of the NPC-NCNC alliance.

According to the Secretary of State.

I am usually optimistic about Nigeria but I confess that these and other recent developments have caused me to share to some extent Mr. Maurice Smith's gloom. It would have been very easy for the Action Group to make friends with the N.P.C. at any time in the past six or seven years. I tried very hard to try to persuade them to do so, and also to back the 'good' Easterners who were then on the Council of Ministers and who broke with Zik (Nwapa Arikpo and Eni Njoku). The Action Group utterly refused, and played straight into Ziks hands when they caused the crisis in 1953—at a time when the North, the West and the good 'East' were all opposed to Zik and his disreputable crowd. The N.P.C. were not blameless in failing to play with the Action Group but they had ample proof of the latter's faithlessness. At that time the Action Group was chauvinistically Yoruba. I do not know whether Awolowo was even then planning for power in the ghout Nigeria. He is now becoming statesmanlike but he is now definitely planning for such power, and will invade Action Group through the East and the North in the Federal elections. I ruefully agree with the last sentence of Mr. Emanuel's minute. J.S.M. 30.7.58 Gloomy—but is there nothing we can do? The Constitutional Conference will be interesting.

With the inevitability of Independence apparent, Zik accepted to go to the Centre as the Governor-General, believing that he would wield the same control over the Armed Forces that the last (colonial) Governor-General did. That turned out to be a miscalculation, although the requirement that the Governor-General had to invite the Prime Minister to form a Government proved to be an asset for him after the messy 1964 elections.

Prime Minister Tafawa Balewa was universally respected, even though he seemed to set a pattern of Nigerian leaders who are good people but are, unfortunately, too weak or lack the vision to change the bad system in which they find themselves. Hence, they can't do much to alter the faulty platforms that propped them to office. These 'good guys' lack the ability, drive or will power to push needed change. Tafawa Balewa, for instance, seemed convinced by the long-standing colonial belief that Nigeria's best chance probably laid with the Action Group running the country but without the political power, as the junior member in a coalition. Also, he apparently desired to have Awolowo as Deputy Prime Minister, although there is little evidence that this would have improved the fundamentally flawed Nigerian state significantly.

Observing Ahmadu Bello outside the perspective of Hausaland's foreign Fulbe hegemony (that also believes it has the guardianship of the Nigerian State) will only fuel images of a disturbing influence and factor on the already faulty and fractured Nigerian Nation.

Nobody, however, deserves to be murdered in cold blood, regardless of your thoughts and perspectives on their person or role.

Bello, and the members of his Fulbe ruling clans were extremely crucial to fulfilling England's colonial objective for Nigeria at the end of the WW2. India's Independence meant the Fulbe control of Hausaland was England's assurance for ground troops in the case a third world war broke out. Indeed, they needed the rest of Nigeria to make the arrangement sustainable, since the ruling clans had 'progressively' destroyed the Hausa economy over their century-long rule.

The English Colonials were, however, very happy to work with Ahmadu Bello in what was effectively a mutually beneficial coalition of foreigners to the detriment of the indigenous countries and peoples of Nigeria.

The mutual need or benefit did not translate to respect however, as official and unofficial communications on Bello show.

'Nigeria': Cabinet memorandum by Mr Lennox-Boyd on the current situation

At the 1957 Conference I agreed to the creation of an office of Federal Prime Minister and to an all-Nigerian Federal Council of Ministers presided over by the Governor-General, whose members would be appointed by the Governor-General on the Prime Minister's recommendation. The United Kingdom interest was secured by the agreement of the Conference that the Governor-General should retain until independence his general reserved powers to act without consulting his Ministers or against their advice, and his responsibility, in his discretion, for external affairs and defence, for the use of and operational control of the Police and for the Federal Public Service.

The leader in the Federal Legislature of the Northern People's Congress, Alhaji Abubakar Tafawa Balewa, became the first Federal Prime Minister and sought to create a 'national' government to work for early independence by including in his team Ministers from the Action Group, who are the dominant party in the West but weak in the Federal Legislature, in addition to members of the National Council of Nigeria and the Cameroons (N.C.N.C.), the major party in the East, and of his own party, the Northern People's Congress (N.P.C.). The Prime Minister is sagacious and able and relations between him and the Governor-General are frank and cordial.

He is openly anti-Communist, he is under no illusions about the difficulties of the task facing both himself and the country, and his policy is likely to be as pro-Western as the narrow Muslim outlook of his principal Northern supporters will allow. (In his party hierarchy he is only deputy to the leader, the vain and pompous Sardauna of Sokoto, Premier of the Northern Region).

Perceptions of Bello by the English outside of their official Colonial circles were probably poorer, as English historian John Iliffe describes.

Accepting with regret the need to participate in party politics, he nevertheless considered democratic competition beneath his dignity, outlawing it in his constituency and leaving interparty bargaining to 'young kids'. He saw politics rather as patronage, appointing ministers without consulting them and treating close civil servants as junior members of his family. Yet qualities deemed 'vain and pompous' by the British won admiration in the north. 'He was always superbly dressed,' it was recalled, 'and the sight of him, and the glittering cortege of large American cars that accompanied him, never failed to please a people that loved display for its own sake.' The crowds delighted in his ostentatious and reckless munificence, which thronged his house with supplicants and threw him into debt, which his bank discreetly cancelled.

There was unfavourable correspondence on Zik. The colonials were apparently willing to take drastic action, including banning and jailing him, but they were worried about the reaction of America and India, as well as the possibility of boosting Zik's image.

The English Colonials deeply despised Zik. The reasons include the American factor and the Ghana Sedition case. More so, Zik did not fit into their plans for Fulbe control of Hausaland. He believed NCNC could win in the West & East, so he was not keen to work with Awolowo, or have Adelabu run as NCNC leader in West, which was a strategy that might have pushed the party over the hump. There was some credibility to the suggestion that he talked and excellent talk but did not have the interest to walk the walk of governance.

'Nigeria': Cabinet memorandum by Mr Lennox-Boyd on the current situation

In the East government has seriously run down. Dr. Azikiwe ('Zik'), leader of the N.C.N.C. remains in power as the personification of Ibo tribalism, but his chronic unwillingness to tolerate around him men of independence of mind has brought into being a Regional Executive Council, almost all of whom are nonentities. Some of the Federal Ministers drawn from his party, notably Dr. Mbadiwe, have recently lost their posts through constant but so far ineffective warfare against his personal dictatorship of party affairs. His star may be on the wane. At present he has no serious rival as leader of the Ibos, who are much the largest race in the Region, and although it would be much healthier for Nigeria if his hand were removed from the helm, the time has passed when H.M. Government could take any effective action to hasten this process. He must be left to the disillusionment of his own people.

The Action Group expected to win in the West, and it did without the falsely assumed cross-carpeting, but English Colonials were terrified about the possibility of Zik taking control of both Southern regions. They considered taking drastic measures to prevent or delay self-government including banning the NCNC.

We also have anecdotal evidence that Hausaland was meant to be the source of troops in case of WW3.

Sir J. Rankine's present view is that the Action Group will win this, but it is always possible that this may be wrong and the N.C.N.C. will emerge in control of both Regions.

The Governor says that if British officers go on self-government, chaos will occur within two years. Many of us are of the opinion that the only feasible way now to stop British officers going on self-government is by political action. I agree that on present form and advice suspension of the constitution is impracticable; but the minimum political action which I believe we can take to prevent chaos is suspension of the promise of regional self-government in 1956. We now have much less time in which to maneuver. The sands are nearly run out.

June 1955

[Eastern Region crisis]: letter (reply) from Sir C Pleass to T B Williamson on the possibility of an administrative breakdown in the Eastern Region. Minutes by M G Smith, T B Williamson, C G Eastwood, Sir T Lloyd and Mr Hopkinson

We should hold back self-government from the East for a minimum of three years from 1956 (though I doubt whether any period should be mentioned even as a minimum). The North would be relieved; and the lesson would not be lost on the West who would be on their best behaviour lest it should 'happen here'. Not everything is against us. The African respects strength. Hence Zik's great respect for Lord Chandos, and his publicly expressed preference for him over Mr. Creech-Jones at a meeting in London after the 1953 conference (with Mr. Fenner Brockway in the chair). Hence too Mr. Awolowo's quiet (if temporary) acceptance of H.M.G.'s decision on Lagos after Lord Chandos had informed him, in reply to his threats of violence, that any attempt to alter H.M.G.'s. decision on Lagos by force would be resisted. 'Black man like strong word'. Zik is a coward. He will be brave to rouse the rabble, but will leave others to deal with the police. He is also a twister. Despite his fair words and perfect manners, he should never be trusted. ***The military and strategic position has changed. Most of the advice given from this Department in the last few years has been based on the assumption that we must at almost all costs have West Africa on our side in the event of a third world war.*** *The latest expert military appreciations show that this thinking is now out of date.* ***Moreover, world prices for many of Nigeria's important exports are falling.***

As the Governor again says Zik must never be made a martyr and unless we could be fairly sure of this the card would probably not be worth playing.

June 1955

[Eastern Region crisis]: letter (reply) from Sir C Pleass to T B Williamson on the possibility of an administrative breakdown in the Eastern Region. Minutes by M G Smith, T B Williamson, C G Eastwood, Sir T Lloyd and Mr Hopkinson

Zik is much too astute a politician to give us good grounds for suspending the Constitution, unless he himself wished us to suspend it, in which case it would clearly not be to our advantage to fall in with his wish. If it were his wish that the Constitution should be suspended, he would of course be delighted if we did that without adequate grounds for our action, since this would give him an opportunity to pose as a martyr not only before his own people but also before America and India;

There is no basis for the Nigerian State outside of English Colonial and Foreign Fulbe interests. Note that the dislike by 'the North' here means the foreign Fulbe rulers of Hausaland. Their word is law in Hausa country, due to the firm control of all the institutions of authority. Indigenous Nigerian communities have had dealings with each other for centuries without the structure of the Nigerian State. Borgu, Nupe and Oyo were probably more connected regions than Oyo and Ijebu, for instance.

Also, nationalities like Igala, the Idoma or Tiv had no fear of Southern domination, and such thought would be just as ridiculous in the current states of Benue, Kogi and Kwara which were gerrymandered into 'the North' by Lugard.

The English also have an overrated view of their performance. Firstly, the colonial adventure in Nigeria was fully transactional, even as the goals and objectives changed over time. The regions had to be profitable, even after accounting for the wages of the colonial officers, hence the need to fraudulently create 'Nigeria' in the first place.

Naturally, not much was achieved. There were very few schools and Hospitals, and there was no mass literacy or development programs outside what was needed to entrench and sustain colonial interests. Hence, the rail lines ran North to South (towards the Coasts) to facilitate the extraction of resources from the country. The focus was on getting resources out of the country, and later 'Military Insurance' for England.

So much more was achieved with less in the decade when there was some sort of competence in Governance in the West, as an example, before that Government was aggressively taken down.

July 1958

'Nigeria': Cabinet memorandum by Mr Lennox-Boyd on the current situation

Educational progress is still slow and there is no prospect for a generation of the Region being able to replace from its own resources the overseas officers who have served them so well. The North fears and dislikes the more educated Southerners and if they were not economically bound to the Federation would be glad to be quit of it.

Reports by M G Smith, A Emanuel, Sir J Macpherson, and Lord Perth, Reporting on the implications of the NPC-NCNC alliance for the elections leading up to Independence.

Dr. Azikiwe chose to go to the Centre as Governor-General (later President), perhaps reluctant to be in an alliance of equals with the Action Group, and believing (or hoping) the role would have command of the Military as it did under the English Governor-General. He was wrong!

July 1958

[Political situation]: minutes by M G Smith, A Emanuel, Sir J Macpherson, and Lord Perth on the implications of the NPC-NCNC alliance

Last month at the Sardauna's invitation Dr. Azikiwe visited him in Kaduna and it appeared at least likely that the visit had led to an alliance between the parties concerned to fight next year's Federal elections not openly as allies but with the minimum of inconvenience to each other and with the object of creating a coalition to exclude the Action Group from power. This alliance was specifically confirmed by Sir Ralph Grey's record below (117) of a talk he had with the Prime Minister. It would seem that the N.P.C. are not confident of their ability to win an outright majority at next year's election and have decided that they must choose one of the other major parties as allies. They have chosen the N.C.N.C., not because they have lost their dislike for it but because internal weakness and division makes it an easier ally than the Action Group. The N.P.C. will at least hope to manage the N.C.N.C. while they might find themselves managed by the big guns of the Action Group. Furthermore, while in the past the N.C.N.C. and the Ibos who support them have been a threat in the North, over the past 12 months the major threat has come from the considerable organising ability and financial resources of the Action Group. This has led to the troubles in Ilorin and to the general Action Group campaign in alliance with the United Middle Belt Congress to win seats throughout the Middle Belt areas. What we have not had confirmed is the statement in the last paragraph of the Manchester Guardian article that Dr. Azikiwe proposed at next year's elections to go to the centre. The inertia he has displayed these last two years would suggest that this is unlikely. But he still might be tempted there by the prospect of the power at the centre that independence will bring. No doubt the N.P.C. will continue to provide the Prime Minister in the person of Abubakar, but the latter's popularity and position in his party is by no means secure (see the Deputy Governor North's letter at (41) on WAF/103/3/01 below). Dr. Azikiwe's crumbling fortunes may well have been restored by this assurance of power following next year's elections and he may be looking forward to taking his place on the African and world stage beside Dr. Nkrumah, if not as Prime Minister at least as Foreign Minister of Nigeria. This might give him just the scope for vague but popular speech-making that he most enjoys, divorced from the detail of internal administration which clearly he dislikes.

On their fears over the soon to be Independent Nigeria and the fact that the Action Group, who they hoped would run the Country, but not have the Political Power were now out in the cold.

July 1958

[Political situation]: minutes by M G Smith, A Emanuel, Sir J Macpherson, and Lord Perth on the implications of the NPC-NCNC alliance

In the Federation the Prime Minister himself has both quality and integrity. Chief Festus (with overseas officers to do the work) makes a bluff and cheerful Minister of Finance. The 2 Action Group members are competent and Dr. Mbadiwe manages to get by. Only in the West is there a team of Ministers with genuine experience and ability able to guide their civil servants and to direct the business of the Region, but these men and their colleagues now at the Centre are to be excluded from Federal office.

There are a very few senior African officers of any value in the East and the best of these, Udoji, has recently been handsomely and publicly victimized by Dr. Azikiwe & has asked to retire with compensation.

However optimistic one tries to be, it is difficult, taking things at best, not to be greatly concerned at the prospect of independence in about two years' time for so large and divided and inexperienced a country as Nigeria. Now that it seems likely that the best brains and experience in Nigeria—the leaders of the Action Group—are to be excluded from power, our concern must be much greater. To apply a famous saying to Nigerian independence—I don't know whether it will frighten the Nigerians, but by Heaven it frightens me.

AZIKIWE, THE NIGERIAN YOUTH MOVEMENT AND PRE-INDEPENDENCE

"Kofo goes to school and all hell breaks loose," might be an appropriate title for an article or satire about the crises that plagued the Nigerian Youth Movement (previously and more appropriately named the *Lagos Youth Movement*) in 1939.

Dr Kofo Abayomi was returning to the UK for graduate studies in Surgery and as such, his position as head of the Nigerian Youth Movement (NYM) and as a Lagos city councilman would be vacant. The movement had surprised all, including itself, earlier when it featured candidates for the Lagos City Council and got elected to three of the positions.

Ernest Ikoli, an Ijaw gentleman was elected as president of the Movement and was expected to take the seat occupied by the outgoing president of the movement, Kofo Abayomi, on the Lagos City Council, in line with its own guidelines. This was however unacceptable to Dr. Nnamdi Azikiwe, editor and publisher of the West African Pilot,. Azikiwe returned to Lagos only two years earlier but had already become an influential member, even if not a ranking member of the movement, so he backed (pushed) Samuel Akinsanya instead to take up the position in the Lagos City Council.

Dr. Nnamdi Azikiwe, Oba Samuel Akinsanya

Samuel (later Oba) Akinsanya was a jolly good fellow, educated, but not to the same degree as most of the leadership of the NYM, but he had major 'street credibility'. He was leader of the transport union (think of a more polished version of MC Oluomo or any of the current Transport Union and park overlords). In addition to this, he was also Organising Secretary of the Nigerian Produce Trader's Union. The impact of commodities in theNigeria of the 30's will make this second role similar to the headship of a combined NUPENG and PENGASSAN (Petroleum Unions) or even the Labour Union.

Zik realized the significant leverage that the "street" could bring, especially combined with his growing media influence, and launched an audacious takeover attempt of the Union through Akinsanya.

Zki was a man in a hurry, and he would not allow a bunch of Tea sipping, English educated socialites stand in his way. Dr Nnamdi Azikiwe, extremely determined to achieve his goals, strived to make the most of whatever situation he found himself. For instance, he was so determined to go to America that he reportedly stowed away on a ship before he was discovered and dropped off in Ghana. He didn't bemoan his fate but took up work in Ghana and was working his way up to save money before he was called back to Nigeria and his father eventually sent him to America.

Zik worked a number of menial jobs and he was a very resourceful person. He attended two historically black colleges (HBCUs) and two Ivy League colleges. This got him exposed to a diverse range of experiences, enhancing his ability to be resourceful. There are not very many people who claim, even today, that they attended two Ivy League colleges and two HBCUs. Added to this fact is that he had lived in many parts of the country: he was born in Zungeru and he studied in Onitsha, Calabar and Lagos. Indeed, Zik was a man of many parts and places.

The fact that he also schooled in America rather than the UK made him one-of-a-kind even among the nationalists. His experience and worldview were very much different from those of not just the leaders of the Nigerian nationalist movement and professionals, but also from those of the colonial authorities and officers. Perhaps, these divergent worldviews brought a certain level of distrust that several people in the Nigerian Independence movement and the Colonials had for Azikiwe. Maybe it was just a lack of understanding, but he also had a bit of a reputation for starting trouble but not being around to deal with the consequences. Maybe it was just a case of his writing being more militant than his personality.

In addition to Zik's recognition of the importance of the streets, media and transport unions to politics, (something a certain Bola Tinubu and other politicians would wake up to several decades later), Zik also seemed determined to stop the election of Ernest Ikoli, a rival journalist, an Ijaw (Izon) gentleman, and non-Afeire[4] to the house because it might have an impact on his own prospects down the line for whatever reason. Samuel Akinsanya left politics all together shortly afterwards and two years later, became the Oba of his own town, Is(h)ara, reigning from 1941 to 1985. So it seemed quite unlikely that he was even much interested in politics. He was noted for referring to the two Action Group rebels Ladoke Akintola and Remi Fani-Kayode as "misguided small boys." They, in turn, reduced his salary to 1 penny per annum.

Zik's return to Lagos from Ghana was hastened by an overturned conviction for sedition for a controversial article published in the African Morning Post. He was the editor of the paper but not the article's author. After he absolved himself of responsibility for the article, his conviction was overturned and he started the West African Pilot in Lagos. The Pilot soon became quite successful and Zik had a conglomerate of newspapers and businesses all over the country within a decade. This media influence made Zik a popular figure in the Nigerian Youth Movement and the ensuing crisis should be viewed in this light.

After failing to get Akinsanya to take Kofo Abayomi's vacated seat, Zik (or as reported, but unlikely, Akinsanya through Zik's newspapers) would bizarrely claim that the Movement (that had just elected an Ijaw as its President, and to the Lagos

---

[4] Author uses Afeire or Afere instead of the popularly accepted slur and exonym for the confederated sub-ethnicities primarily located in the Southwest of Nigeria and their related ethnic nationalities.

Council) was discriminating against Ijebu-Afeires, and his own ethnic Igbos. This led to a split in the movement, with Zik taking most of the Igbo members out of the movement, instantly gaining a platform and base to form a partnership with the aging Herbert Macaulay, who had been knocked off the political scene by the NYM, a few years later.

It was not the last time that Zik would leave a trail of confusion when not able to have his way. Obafemi Awolowo, an Ijebu-Afeire himself, was clearly not impressed with Zik, or his methods.

For most of the 30s, the NYM, which was more of a league of educated African professionals and intellectuals in the city, got on very well with the English Governor-General Bourdillon, who believed that nationalism was understandable, but needed to be managed. Bourdillon was somewhat impressed by the professional cadre of the members of the movement and the surprising gains they had pulled in the Council elections. Thus, he was quite friendly with the ranking members; Azikiwe even became his tennis partner.

Perhaps due to his friendliness with Bourdillon, Azikiwe and members of the youth movement supported the English war effort in World War II, especially since there were unfounded rumours that Nigeria was to be ceded to the Nazis either if England lost or by the British to appease Nazi Germany. Zik and six other editors were invited to the UK in 1943 on a trip sponsored by the British council to view life in England in general and especially during the war.

The British government had a chaperone to accompany the journalists, a colonial information officer and he documented his thoughts on Zik. He thought that Zik was sincere in his opinions but unscrupulous in his way of getting them across. He felt that Zik was a very clever man with excellent manners, but with a penchant for skating on thin ice. it appeared that Zik was not trusted by the other editors. Again, this might be a reflection of the English perception of the American background and personality in general.

Bernard Bourdillon retired shortly afterwards, followed by John Shuckburugh and Alan Burns, men who were said to have a mindset of subduing nationalism. This was probably in line with the English Colonial objective of getting a couple of million troops out of Hausa country if there was ever a third World War.

The hardening of the colonial grip however got Zik more frustrated and his writing became more militant rather than conciliatory and supportive of the colonials as it had been during WWII.

It was said around this time that Zik's writing helped radicalise, at least, from the English perspective, the nationalist movement, with the Zikist movement and its revolutionary undertones, emerging, and the NCNC started pushing for constitutional reforms. Zik and the NCNC wrongly assumed that Nigeria was not a league of forced countries, and opposed the rights of regions to secede.

Zik supported the general strike of 1945, while Ladoke Akintola, a journalist and former teacher at the Baptist Academy, opposed the strike. Some suggest Akintola was used by the English Colonials for their interests and agenda throughout his time on the Nigerian political scene, and this narrative gained traction and credibility as the 36 year old Akintola, already a leading journalist, got an English scholarship to study law in 1946.

This was the basis of the rivalry and suspicion between Akintola and the young Anthony Enahoro who was editor of Nnamdi Azikiwe's paper, the Southern Nigerian Defender. The paper was established in Warri, but its office was later moved to Ibadan in 1944.

Zik and the colonials both had a trust and timing problem. Regarding the timing, the 1940s and 1950s just seemed to drag on forever for Zik who believed that he ticked all the boxes to be the post-war and post-independence leader of Nigeria. But the English definitely had issues, at least, in the 40s and early 50s with an independent Nigeria (due the 'Insurance policy' that Hausa Country provided, with the reality of an Independent India). They suggested that it would take a generation, not the maximum of 15 years after the war that Zik had suggested.

Even though the colonials did not think that Zik would resort to subversive or revolutionary activities or militancy, they were worried enough about his writing and ventures and decided that neither he, nor the NCNC would play a part in their plans and arrangements for Nigeria. This explains the continuous pressure on Awolowo and the Action Group to go into an alliance with the NPC, a suggestion that Awolowo consistently rejected for over a decade. It seemed Akintola and Fani-Kayode were, however, a tiny but significant minority in the AG leadership in this regard.

Zik's pathway to Nigeria's leadership was straightforward and simple. He had the East on lockdown and only needed to pull an upset in the West; he did not need an alliance, and had been strategizing for this since the orchestrated NYM split of the 30's. Awolowo and the Action Group had a different approach, wanting to go into an alliance on its own strength and not as a junior partner. Hence, it needed to secure the West and form alliances with other indigenous Nigerians that it felt were under the oppression of foreign Fuduyan Fulbes.

Whether real or imagined, the colonials also felt that Zik was dictatorial and also poor with administration and details, which is hard to believe for a former editor. Maybe Zik had just become tired of waiting for what he thought was his, after two and half decades of activism without independence. The colonials also believed that they had enough in terms of infractions to jail him and proscribe the NCNC. However, they were afraid of international reactions, especially from the USA where Zik had a large base of friends and support as well as the newly independent India. Thus, they decided not to arrest him or proscribe the NCNC, believing that if they were able to push through the Action Group and NPC alliance, Zik would naturally fade into obscurity.

With this background in mind, it was most surprising to most observers, including the English Colonials, that Zik pulled the joker and went into an alliance with the Ahmadu Bello led Northern People's Congress for the 1959 elections leading up to Nigeria's independence. It appeared that Zik assumed that his NCNC would have equal powers with the NPC under this arrangement. He also thought that, as Governor-General, he would have control of the military, just as the outgoing English Governor-General did. This, however, proved to be a gross miscalculation, despite Zik's attempt to force the position of the Governor-General and later President into that of a Commander-in Chief.

The limitations of the powers of the office proved to be a source of great frustration for Dr Nnamdi Azikiwe. This played a huge part as things blew up and eventually it degenerated into the Nigeria-Biafra civil war.

AKINTOLA AND THE FAILED ALLIANCE BETWEEN NPC AND AG

There may have been markers from the past to suggest the controversial role that Akintola was going to play in the Action

Group's (and eventually Nigeria's) crisis. However, there was no indication from Remi Fani-Kayode's background of the fact that he would be Akintola's co-traveller in that process.

Born in England in 1921, Remi Fani-Kayode had completed his law degree by his early 20s. He went on to co-found the law firm Thomas, Williams & Kayode with Bode Thomas and Rotimi Williams. Boasting a privileged background, he had no reason to be intimidated by or beholden to the English colonials. If anything at all, like Bode Thomas, Awolowo, Enahoro, Jakande, Chike Obi and the rest of Action Group stalwarts, he was expected to combine his education and cultural pride to pursue home-grown solutions and aggressive development policies.

However, like Akintola, he was also expected to know that the English colonials trying to force an alliance between the Action Group and the NPC could not have been for the good of their people or region.

After the 1966 coups, he withdrew from Nigeria and the political scene for some time, but would re-emerge late in the 1970s, serving as the vice-chairman of the Fulbe-led National Party of Nigeria that produced President Shehu Shagari. He also famously formed a law firm with Sobo Sowemimo, the brother of the judge that jailed the Action Group leadership (about 27 of them) in 1963. The Judge, George Sowemimo, was appointed as the Chief Justice of Nigeria, by President Shagari in 1983.

Remi Fani-Kayode and Shehu Shagari

Fani-Kayode's motivation for his actions and decisions can hardly be explained by a quest for political ascendancy or wealth. Perhaps some undisclosed issues the now-defunct Thomas, Williams & Kayode or a hold by the English colonials that hasn't come to light may be responsible.

To be fair, in the 1960 - 1965 period, each of the major parties was deeply interested in not just consolidating its base, but grabbing new frontiers. The Action Group, which had spent the previous decade refusing colonial suggestions to merge or ally with the NPC, certainly felt slighted that significant parts of Afeire Country and affiliated ethnicities, including Ilorin, Kabba, Offa, and Igala Country were gerrymandered to the 'North'. Members of the party felt that with some effort, these places could be regained and reinstated (at least politically) into the Western Region. They probably also felt that there was no basis for the foreign Fulbe control of neighbouring indigenous areas, which was what Ahmadu Bello's leadership of the NPC represented. This drove the partnership with Joseph Tarka and the United Middle Belt Congress.

Also, the AG leaders believed in true federalism and ideologically didn't consider a strong alliance necessary. However, the party expressed its willingness to enter into an alliance with Nnamdi Azikiwe's NCNC if it became necessary. Thus, it seemed that the AG leaders were not entirely opposed to forging alliances with other parties; they only wanted to do so from a position of strength.

On the other hand, Azikiwe thought he and his party could go solo and gain a majority of federal seats if only they could win the Western region. Such an assumption was not as farfetched as it seemed. Azikiwe had the power of the media and he believed that by forming strategic partnerships with influential figures (which it didn't do, with great vigor or much success), the NCNC could defeat the Action Group in the west while retaining its Eastern region domination.

Already, the circumstances of the Action Group's emergence was already a problem for Azikiwe. The party materialised out of the socio-cultural group, Egbe Afenifere in 1951. Before then, Azikiwe was confident that the NYM palava (caused by him) was only a slight bump on his path to the leadership of a Nigerian state that would be Africa's most populous. Maybe it was even a statement of intent. This explains his irritation with the Action Group, especially since there was virtually no threat to his hold on the East; the West was the problem.

On his part, Ahmadu Bello believed that with English Colonial support, he could dominate the Nigerian State. He was thus open to the Colonial Government's proposal of being the senior partner in an alliance with the Action Group as this would not only consolidate the hold of his foreign Fulbe hegemony over the already subjugated Hausa country, but could instantly increase the controlling sphere to all of the countries in the Nigerian State.

Bello never got over losing an attempt at the leadership of the Fuduyan Fulbe Clans to his relative Siddique Abubakar, but embraced the means of the colonial Nigerian State as a means to achieve his objectives, following the colonial rescue from an attempt by his clan leader and relative, to jail him over real or imagined infractions having to do with cattle.

Bello, soon assumed political leadership of Hausa country, and the gerrymandered North through the NPC, and also tried to create an alternative religious authority to clip the wings of his rival and clan leader, Siddiq Abubakar, in partnership with Fulbe religious scholar, Abubakar Gumi.

Bello often used shocking language (and seemed determined to back it up, with the TIV situation and the Action Group crisis being examples) pertaining to slaves, swords and even conquered Hausa Country, and clearly identified his clan as foreigners, who similar to (or in partnership with) the English Colonials "owned" the emerging Nigerian State.

The world shifted a bit between 1950 and 1965. Fears of a third world war soon gave way to the Cold War. The English colonials, historically and institutionally, maintained strong support for the NPC and Ahmadu Bello, but this was based on mutual self-interest.

It was therefore not surprising to experienced observers that as the first military coup happened, the British declined to intervene militarily (or even acknowledge the request from the scrambling NPC politicians), choosing instead to sit back and watch where the pendulum of power would swing; whoever emerged winner would have English support.

After all, their policy in the earlier part of the previous 15 years had been to ensure that they could get Hausa troops to fight for the English cause. They were definitely not prepared to have British soldiers and lives lost for the Fuduyan Fulbe cause in Nigeria. That would be the complete opposite of their objectives, plans, and strategy.

Gen. Aguiyi-Ironsi swung the pendulum to himself and seized control amidst the confusion of the January 1966 Coup. This was a move with bad intentions, that was equally badly managed, and it expectedly had very bad consequences.

IT TAKES TWO TO BRING THE HOUSE DOWN

Ladoke Akintola was very different from the younger partners from privileged backgrounds in the law firm Thomas, Williams & Kayode. Their parents could afford an education at the best universities in the UK and the young men returned to Nigeria in their 20s to set up a law firm. On the other hand, Akintola had to 'earn' everything the hard way. He could only attend law school after he had crossed over to journalism after a long teaching career at the Baptist Academy.

As a journalist, he had opposed the Azikiwe and the NCNC strike in 1945. That also put him in direct opposition with the young Anthony Enahoro, who was the editor of Zik's newspaper, and who continued to view Akintola with suspicion and distrust, even when they were ranking members in the Action Group. Akintola and Ahmadu Bello had a particular dislike for Enahoro, and the then deceased Bode Thomas, and put pressure on the English Colonials to extradite Enahoro, who had escaped the mass crackdown on the Action Group Leadership through Ghana, before eventually ending up in England.

After Akintola's English-funded law scholarship, he returned to Nigeria and worked under the younger partners before crossing over to link up with Dr Chris Ogunbanjo in another law firm. He also dipped his feet into politics, initially as a legal adviser to the Action Group.

Perhaps all these young men would have lived much wealthier and less stressful lives if they had just remained in private law practice. Rotimi Williams became one of Nigeria's foremost lawyers, with a glorious career that spanned over 6 decades. He was to Nigerian law what his brother, Akintola Williams was to Nigerian accounting: a doyen. Chris Ogunbanjo had a similar impact in law (and remained one of Nigeria's top lawyers into his 90's.

Rightly or wrongly, these young men were attracted by politics, perhaps motivated by a desire to help build the emerging nation, or just a self-serving desire to be the ones 'running things'

Akintola's supporters have tried to play down his part in this crisis that soon overran the Action Group, and the rest of Nigeria, by suggesting that having grown up in Minna, on the border of Nupe and Hausa countries, his perspective was a bit different from the rest of his party stalwarts. The fact that he spoke Hausa fluently (he also spoke Nupe) meant that he could relate with Ahmadu Bello a bit more intimately (the Fulbe ruling class adopted Hausa as their secondary language in addition to their native Fulfulde). Akintola, reportedly, helped bring Ahmadu Bello and co back to the table following the continual walk-outs staged every time the Action Group raised a motion for Nigeria's independence.

Maybe he deserves some credit in that regard, but it probably would have made sense if questions about Nigeria's future and the differences between the various constituent indigenous countries had been emphasized and settled at that point. Instead, bringing everyone back only meant making collective decisions that served nobody's interest at the end of the day.

Following the death of Bode Thomas at the young age of 34. Akintola and the Alaafin (Adeyemi) decided to symbolically dance on his grave with their actions afterwards. Bode Thomas, Lagos-born, held the high non-royal title of *Balogun* of Oyo and shortly after his death, the Alaafin Adeyemi II appointed Akintola as the Aare Ona Kakanfo, which was not an Oyo title but one that represented the traditional field marshal of the entire Afeire Confederacy. The office had been unoccupied for about 80 years since Latoosa of Ibadan.

In addition, Akintola contested to replace the late Bode Thomas as deputy leader of the Action Group, and despite Awolowo's obvious preference for Chief Arthur Prest, Akintola had his way with the strong support of Fani-Kayode.

Akintola and Fani-Kayode may have reasoned that their actions could reverse the fortunes of the party following its loss in the federal elections of 1953, which was due to the unpopular taxes that were introduced to finance free primary Education and Healthcare. The Alaafin was particularly opposed to this tax, both because he felt it increased the burden on his people, and limited the role of his Native Authority. Bode Thomas, Awolowo's Man Friday, and major proponent of the party policies, was now at odds with his Alaafin, who was openly campaigning for the opposition NCNC.

L: Fani-Kayode with Akintola R: Bode Thomas, Awolowo and Akintola

However, it soon became obvious that the taxes were being put to good use as the region had shot ahead of other parts of the country in development, education, and free healthcare for children. Within a couple of years, the party's popularity was sky-high and talk about bringing the region closer to the centre made no sense. Awolowo used the popularity of the party, to depose the Alaafin, an act that would have been considered sacrilegious at any other time, but which passed without much of a whimper from the public, due to the popularity of the party. This did not end the problems within the party though.

As the legal adviser of the party, Akintola certainly knew about the resistance of the likes of Awolowo and Enahoro to the British push for an alliance between the Action Group and the NPC. So, Akintola's continued dalliance with Ahmadu Bello was nothing short of defiance to, and open rebellion against his own party's leadership, agenda and philosophy. Perhaps the English Colonials had something on him or he was just carrying a moral burden to pay for his scholarship.

Despite all these, Awolowo decided to give up the premiership of the Western Region to serve as the leader of the opposition after the 1959 elections. It was a curious decision that may have proven very costly for the emerging Nigerian nation.

Despite the clear unpopularity of this move, Akintola's open romance with the Fulbe NPC leader, Bello continued. It got to a head when he was absent from his own party's convention in 1962; instead, he was with Ahmadu Bello at the University of Ibadan commissioning a hall in honour of Muhammed Bello, the second Fulbe ruler of Hausaland, slave raider of millions of indigenius peoples, and the younger Bello's progenitor. This provocation led to the removal of Akintola as the deputy leader of the party, which automatically meant impeachment from his position as the Premier of the Western Region. Ahmadu Bello quickly expressed a very strong and public support for Akintola.

There was a very minor fracas that was limited to about five NCNC members in the Western region legislative house when it was time to ratify Akintola's removal as premier. These four or five NCNC supporters of Akintola in the 80-member house tried to disrupt proceedings by seizing the mace, and based on that, the NPC-led federal government unbelievably declared a state of emergency in the Western region. This definitely had Ahmadu Bello written all over it, and sadly despite constant complaints about the excesses of his party leader, Prime Minister Tafewa-Balewa once again allowed himself to be a pass-through, and failed to show independent leadership at this extremely pivotal point in Nigeria's steady descent to chaos.

Later in 1963, with Akintola reinstalled as Premier, the entire Action Group leadership was rounded up and tried, alongside their strategic allies such as Joseph Tarka of the United Middle Belt Congress. Anthony Enahoro was able to escape to Ghana and then England but eventually, despite some controversy, he was extradited to Nigeria to face trial and jailed along with most of the leadership of the Action Group.

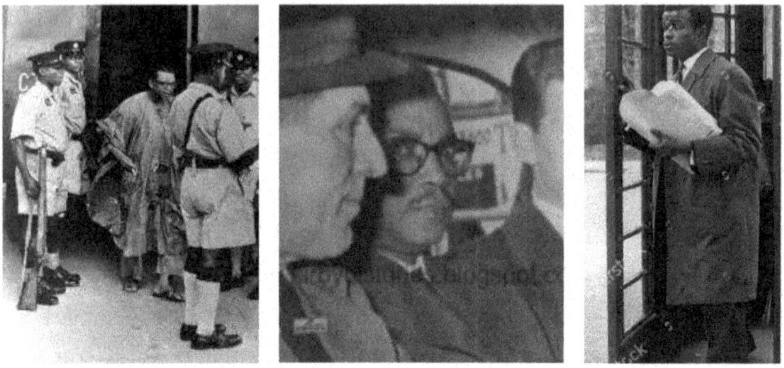

L: Chief Anthony Enahoro back home for trial C:Enahoro in English police custody R: A young Abraham Adesanya was on the legal team fighting against the extradition of Enahoro

The NPC-led federal government eventually reinstalled Akintola to the Premiership of the Western Region despite the Action Group's election of Dauda Adegbenro as his replacement.

Even with significant external support and almost all the leadership of the Action Group behind bars, Akintoa did not get the support and popularity from the public that he wanted, and maybe even craved. The 1965 elections that he contested were so significantly rigged that the head of the electoral commission refused to declare him as Premier and rather resigned from office. Adegbenro and the Action Group declared themselves as winners and the crisis in the Western Region, alongside other controversies going on with Zik and Balewa in the NCNC-NPC governing coalition in the Central Government (which included both men trying to assert control over the Military), led to the environments in which the the military coups of 1966 and the Nigeria - Biafra War from 1967 occurred.

1966 COUPS - DESCENT TO CHAOS

The Emmanuel Ifeajuna-led coup of January 1966, and the Murtala Mohammed led attempt to divide the Nigerian State in July 1966, have been the subject of much discourse since then, with the former getting a larger share of the blame for triggering the subsequent civil war and every other major problem Nigeria has faced since then, than the latter, which, included a brutal slaughter of almost 300 non political regular officers and soldiers and thousands of civilians.

Considering the short and straightforward history of the Nigerian State as we have done, isolating the events of 1966 is absolute hogwash, and amounts to a continuous attempt to avoid the obvious fact.

The very problem of Nigeria is Nigeria. The nation was set up by foreigners, for the benefit of foreigners and for the plundering of the local countries without the consent or input of the local peoples. This core essence and raison d'etre of the Nation can almost certainly not be changed.

This fundamental deformity and flaw is forcefully suppressed, but never addressed, and those in control of the State pretend the foundational deficiencies don't exist and make vain but costly efforts to extinguish the ability of the constituent countries (wholly or partly) in the Nigerian State to function and progress. Little wonder that the Nigerian State has become the world's poverty capital, has the most out of school children, and is one of the worst places to live. The foreigners who are there to plunder, keep doing so aggressively, with a zombie-focussed determination to feed on the corpse of the Nigerian nation, and its indigenous countries, if necessary.

One cannot help but see the obvious disconnect, especially since the indigenous countries and ethnicities have so much going for them in terms of resources, culture, and hardworking and enterprising people, and have always been sustainable. The problem with Nigeria is and has always been Nigeria.

The first two military coups will be examined as one continuous event, with the necessary detail provided for the July 1966 coup, unlike other accounts that gloss over these events, perhaps because many of those mutineers ended up being the rulers and shapers of what exists of the Nigerian state (as at 2021). No one is known to have been officially reprimanded, detained, interrogated or punished for sponsoring or executing the events of July 1966.

Most of the conspirators of the January 1966 attempt, though not yet officially punished, died shortly afterwards.

There is some pretty solid evidence that Nigeria's first President, Dr. Azikiwe, either bluffed about a coup, or was approached about one, as the crisis in the NPC-NCNC ruling coalition got worse in 1965. Given Zik's tendencies, he may have been bluffing, but there seems to be evidence of discussions of a coup with Ojukwu, who then tried to bring Yakubu Gowon, David Ejoor and Victor Banjo on board. Gowon and Ejoor outrightly rejected the idea[5], but Banjo didn't, and may have been waiting to be carried along by the others. That double-mindedness was a mark of his tragic appearance on the stage of Nigerian history.

The fact that Lt. Col. Odumegwu Ojukwu theorized or advised that a coup that is perceived to be ethnically biased would be a colossal disaster, gives the impression that he was already aware that the Ifeajuna plan was in the works.

While Azikiwe and Tafawa Balewa clashed over censuses, elections, election do-overs, and who had control of the military, Major Emmanuel Ifeajuna, Nigeria's first Commonwealth gold medalist, had started making plans for a coup. Ifeajuna was at the center of at least two groups with interests or roles in the coup, or to whom notification was given about the coup.

His fellow conspiring junior officers believed they were trying to instigate a revolution against the political class and the senior officers. On the other side, it seemed, were Zik and Aguiyi-Ironsi, who possibly individually sanctioned the coup or, at the very least, been notified ahead of the events. Like Murtala Muhammed would react to Yar'Adua a decade later, both Zik and Aguiyi-Ironsi may have given the impression that "well, I am [6]not against it, but don't mention my name if you get caught, and we'll do everything to save you." That was probably their stance towards the planned coup.

Ahead of the coup, Zik departed Nigeria on a medical vacation, again there was a parallel to the actions of Murtala Muhammed a decade later, who also, somehow, was outside the country while Shehu Musa Yar'Adua plotted Yakubu Gowon's removal. Zik's personal Doctor who accompanied him on the

---

[5] https://www.vanguardngr.com/2020/01/ojukwu-banjo-conceived-what-could-have-been-first-coup-in-1964-gowon/

[6] https://thenewsnigeria.com.ng/2021/01/15/55-years-of-nigerias-1st-coup-what-british-intelligence-officials-wrote-about-it/

trip, eventually abandoned him abroad in December 1965, when he could not figure out the need to extend the trip.

That first coup eventually occurred on January 15 1966, with Kaduna, Ibadan, and Lagos as the primary action points. Prime Minister Abubaker Tafawa-Balewa, Finance Minister Festus Okotie-Eboh, Northern Region Premier Ahmadu Bello, and his wife Hafsatu, Brigadier Ademulegun and his wife Latifat, Brigadier Maimalari, Colonel Shodiende, Colonel Kur Mohammed and Lt.Cols Unegbe, Pam and Lagerma were among the leading politicians and senior military officers murdered by the Coup plotters. About 10 other civilians and military officers were also killed during the Coup. The killing of Lt. Col Unegbe, a senior Military supply Officer, with no key command or control over troops, but significantly the only Igbo officer killed, only seemed to raise suspicion of a senseless murder done to give the impression of non-ethnic agenda[7].

L:Brigadier Maimalari R: Brigadier Ademulegun: Maimalari was Nigeria's first Regular Officer. Not promoted from the NCO ranks, or with a Short Service Commission, or a Subaltern or other contract training arrangement.

7 Some suggestion of a personal dispute stemming from a love triangle has been stated as the reason for Lt. Col. Unegbe's murder.

At the time, Major Olusegun Obasanjo had just returned to Kaduna on a course and was staying with Major Nzeogwu; he did not play any role in the coup. However, another officer who had recently returned from an overseas course, Yakubu Gowon, would play a crucial role in suppressing the coup. Indeed, he was one of the few shining lights amidst the tragedies of both military coups. Other significant participants in stopping the coup, or limiting the carnage were Lt. Col. David Ejoor and RSM Tayo of Bonny Camp.

David Ejoor was the commander at Enugu; he happened to be in Lagos at the time. According to reports, he either unilaterally arrested the convoy coming from Abeokuta or at least decoyed the group towards their arrest/surrender. Then he returned to Enugu to check the situation at his base.

It was RSM Tayo of Bonny camp who, realising that something was amiss, locked the armoury and had all the NCOs left in the barracks to sit in the open field. Since there was no precise information on who was attacking, he took the measure to forestall violent incidents out of irrational fear.

The principal executors of the coup were Nzeogwu in Kaduna, Nwobosi in Ibadan, Ifeajuna, and others, including Ademoyega in Lagos. Nwobosi arrested the deputy Western premier Remi Fani-Kayode in Ibadan and brought him to Lagos, where he (Fani-Kayode) was eventually released by Gowon and reportedly taken away by Victor Banjo afterwards.

Even to the most liberal mind, General Aguiyi-Ironsi's actions relating to the coup gave room to doubt that he was ignorant of the events that unfolded on that day. First, he supposedly received a phone call from Lt. Col James Pam, informing him of the coup, around 3 am on the morning of the coup. Pam, however, did not live to give an independent corroboration of this fact. Ademoyega had led a team to disconnect the manual

Lagos telephone switch around this time. Several key personalities, including the British high commissioner, could not use their phones that morning, but it seemed Aguiyi-Ironsi did get the call that tipped him of the coup.

Aguiyi-Ironsi claimed to have intimidated the boys on Carter Bridge led by Captain Orji to cross over from Lagos Island to the Mainland on his way to Ikeja Cantonment. This was strange given the violence perpetrated by the plotters a few hours before. Aguiyi-Ironsi eventually made it to the military cantonment in Ikeja on the mainland of Lagos, close to the airport. It seemed he and the brigade commander just got the officers together and waited without any specific action being taken. Gowon, who was the incoming commander of the base and had just returned from a lieutenant colonel's course in the UK a few days earlier, was the one who mobilised a quick reaction force to the island of Lagos, where all the government offices and key leaders of the country resided. Gowon's quick reaction force included a certain Lieutenant Muhammadu Buhari, and the team was able to check on all the key senior officers and get a real-time assessment of who was missing. They also arrested a couple of the plotters and executioners.

In the early morning hours following the coup, it became apparent that the Prime Minister was missing. The most senior NPC minister, Inuwa Wada, was in Europe for medical treatment as the coup played out, leaving Bukar Dipcharima, a former Zikist who was previously in the NYM and NCNC as the most senior NPC politician in the cabinet. He had approached the British High Commissioner Francis Cumming-Bruce for possible military assistance, but it seems Cumming-Bruce, after consulting with some other British officials locally, decided to sit back and see which side emerged victorious from the situation and support whoever won the control of the Nigerian State.

As previously mentioned, English interests in Nigeria after the Second World War centred around having access to Nigerian, especially Hausa troops to fight and die for the English cause if necessary. Of course, there was no appeal to having English soldiers come to die for and in a Nigeria that had lost its strategic importance. This was just before the emergence of Crude Oil as a major strategic interest.

In any case, the Senate President Nwafor Orizu, as Acting President (in Nnamdi Azikiwe's absence), refused to swear Dipcharima in as the interim Prime Minister. The NCNC ministers insisted that Dr. Ozumba Mbadiwe was the most senior cabinet minister and should be the acting Prime Minister instead. However, the NPC insisted on Dipcharima. It was important to have an Acting Prime Minister, because the British High Commissioner maintained that a request for military assistance had to be channeled appropriately through a serving or acting Prime Minister. Nevertheless, it didn't seem that the British were going to grant that request.

Perhaps disgusted by the inability of the politicians to choose an acting Prime Minister or just trying to seize the chance to rule the country, Aguiyi-Ironsi had a meeting with his junior officers, supposedly trying to get the support of the military to take over. Gowon, again, was a strong voice in opposition to the military getting involved in civilian rule. Aguiyi-Ironsi also met with a gathering of the remnants of the government comprising cabinet members and other senior politicians. However, he supposedly called Orizu aside for a private meeting; once Orizu came out of the meeting, he declared that he would not swear in Dipcharima as acting Prime Minister. Eventually, the politicians yielded power to Major General Aguiyi-Ironsi as the civilian cabinet members were forced to read a prepared speech with gun-toting soldiers on standby. Cabinet members reported that he also gave them the 'You either hand over as Gentlemen, or you hand over by Force' option.

That was a major mistake on the part of Aguiyi-Ironsi, and the allure of power may have shielded him from the reality that a direct Military Government, especially under these circumstances was only going to increase the instability inherited from the political process, and not quell it, especially if he was going to be under any constraint, time or otherwise, regarding the quick trial of the January 15 plotters.

David Ejoor, possibly one of two officers leading efforts to actively investigate and suppress the coup, gave a troubling account of Aguiyi-Ironsi in Ikeja, pistol drawn, and asking him, David, are you with me or against me?

Victor Banjo again played a puzzling role in Nigeria's history, as he was quickly arrested afterwards during a visit to Aguiyi-Ironsi's office, where he had suggested that he was ready to help the government. He may have been trying to gain some leverage on his wavering 'yes' to Ojukwu's request a few months earlier, but as these things play out, there is never enough room at the top and the Governorship of the Western region (which would have matched Banjo's 'rank and profile') was already penned down for Adekunle Fajuyi, who was sympathetic to the Ifeajuna coup.

Aguiyi-Ironsi's best option was probably not to accept or take over power in the first place, but once he was in power by happenstance, opportunity or design, his best option at survival and stability in the country, outside of a quick trial of the coupists, would have been to separate the military along regional lines and tried to solidify federalism or aim for a loose confederation. That would have diffused the tension to a large extent and communicated that there was no plan of domination by one group or ethnicity over the rest of the country or another. The fact that there were real imbalances in the numbers (Hausa dominated NCO rank and file for instance) would not have been a problem if a regional government was increasing the numbers for its regional Army. The Bello-Ribadu policy would have been no problem for a regional Military.

Rather, Ironsi seemed to pursue a Unification approach, and appointed a Commissioner for an Administration for a Unified Nigeria, while abolishing the Hausa language requirement for the Northern Civil Service. This, along with a speech by the Eastern Military Governor, Lt.Col Ojukwu that Eastern Civil Servants could be transferred to other regions based on seniority only increased fears, especially in the far North that the January 1966 coup was a means for subjugation.

Aguiyi-Ironsi also delayed the trial of the coup plotters. Some have suggested that he was afraid of putting Ifeajuna on trial. This was highly suspicious, especially since other separatists who were not coup plotters like Isaac Adaka Boro and his Niger Delta Volunteer Force colleagues had been arrested, tried and sentenced in that same period. It was indeed curious that Aguiyi-Ironsi, who had lost two of his Brigadiers, and had claimed to be top of the elimination list of the January plotters, was viewed as being personally responsible for delaying the trial of the coup plotters. Ademoyega wrote about Ifreajuna's opposition to having him handle this critical part of their plot, and assigning him to the telephone disconnection, and other assignments on the day.

In addition to Aguiyi-Ironsi's strategic and governance errors, there were other basic tactical and operational mistakes that also led to his death in July 1966.

The plotters of July '66 knew that Lagos was Aguiyi-Ironsi's stronghold and that an attack there was likely to meet a lot of opposition. So, they had planned to carry out the coup outside of Lagos. Iit was an operational failure therefore, that there was no immediate evacuation plan (perhaps with a chopper) in place for the Head of State in Ibadan and for his shuttles to places outside of Lagos. Ironsi requested for a helicopter, which arrived after he and Fajuyi had been led away. In any case, it could be argued that a chopper would have been an easy target with the building already surrounded. It might have been a deterrent, if already part of the Head of State's entourage though.

After the first coup, especially with the delay in trying the detained coupists, Lt. Col. Murtala Muhammed started coordinating action for a counter-coup, setting a date for July 29. He called off this initial date, fearing that his plot had been discovered when he was confronted by an intelligence officer, Lt. Col. Patrick Anwunah, who said he knew the plan was to kick off that night. Anwunah notified a couple of commanders, including the garrison commander in Abeokuta, Lt. Col. Okonweze. The latter gathered all the officers in his unit together in the mess and tried to appeal for calm and reason to prevent any bloodshed.

This peace meeting, however, in a counter-productive fashion, triggered the July 1966 coup. One of the NCOs assumed that there were plots to eliminate Northern soldiers. This lone NCO rallied his colleagues, and the Armourer, who issued weapons to northern NCOs. These NCOs promptly went into the meeting conveyed by their Commander and killed him, and other officers of Igbo origin.

It didn't help that Lieutenant Mwadkon Pam, the younger brother of Lt. Col. James Pam, who was killed in January '66, was serving in Abeokuta and was out for revenge. Okonweze had planned to take all soldiers to the Armoury to pick up arms and ammunition so they could defend themselves. A radically different strategy to that taken by the Warrant Officer in Bonny Camp six months earlier There is no guarantee of success by either one of these strategies however. The Okonweze strategy of issuing weapons to act as a deterrent, was successful in Enugu, for instance.

Following the killings in Abeokuta, Lt. Mwadkon Pam placed a call to Lt. Garba Dada to inform him that 'Aure Paiko' was in full gear. Dada then rallied the conspiring Soldiers of the 4th battalion in Ibadan (seen as one of the successor Divisions of the 'Glover's Hausa's, the Army's founding unit), and woke the visiting Major Theophilus Danjuma, who was on the entourage of the Head of State, but was staying in the barracks.

OPERATION ARABA/AURE PAIKO

When the mutineers in Abeokuta contacted Murtala Mohammed, it reportedly had a domino effect on most or all the other army formations, with the feeling that failure to follow through would leave the guys in Abeokuta exposed. Secondly, failing to arrest the head of state, Aguiyi-Ironsi, would also make the efforts fruitless. With Murtala Muhammed as the unofficial command centre, the killings started at different military formations.

For most Nigerians born in the 70s and 80s, the official accounts and records of contemporary Nigerian history, especially the events of the 1960's and 1970's are scant and essentially useless.

There are however pointers and indicators that force an extremely positive bias or impression on some of the actors and participants in the major events, and perhaps no one is a bigger beneficiary of this, than Murtala Ramat Muhammad, His face was on the 20 naira note, the highest local currency (fiat money) available at the time and for a couple of decades. His name adorned the leading international airport in the country,

and he is meant to occupy an exalted status in the minds and lives of Nigerians, a combination of Kennedy, Eisenhower, Roosevelt, and General
Douglas MacArthur all rolled into one.

Many Nigerian kids at the time grew up knowing the rhyme disparaging the 'wife' of Buka Suka Dimka, who was accused of killing Murtala Muhammed. It was something along the lines of, "Dimka's wife, what are you doing in the market? Shame (on you); Go home because Dimka is about to be executed." So, as a child growing up then, in your mind or your subconscious, Murtala Mohammed occupied a saintly, sacred role that should not be questioned.

Sir Ahmadu Bello, with his Officers at Mons.

The reality is not quite straightforward. Murtala Mohammed may have had a change of heart as he became the head of state, primarily due to the events that led to his assumption of the position, which occurred about a decade after the current reference point. However, the pre-1975 Murtala fell far short of the image presented by the scant official records and propaganda.

The entire July Coup/Partition attempt was nicknamed *Aure Paiko* or 'Paiko's wedding'. 'Paiko' was the nickname of Lieutenant Garba Dada, the adjutant of the fourth battalion in Ibadan, Perhaps Dada was to get married around the time of the coup, which would provide a cover for the plotters if heard discussing the upcoming 'wedding' in Ibadan.

Captains Bali and Remawa in Abeokuta were the first to notify the Army headquarters of the killings. Both were out of the barracks on a night out to town and thus were not part of the action, but when they got back to base, they saw the dead bodies, and got in touch with Army HQ in Lagos.

Lt. Col. Gowon, now the Army Chief of Staff, had to worry about suppressing a rebellion for the second time in six months. This time, he faced a formidable opponent in Murtala Muhammed, who had the backing of the majority of the Hausa and other Northern NCOs of the Army and the entire Bello-Ribadu Officer Corps, which was now a significant portion, if not the majority of Junior Officers (the Second Lieutenants, and Lieutenants especially) of the Army.

It appears that by 3 am, soldiers from the 4th battalion led by Lieutenant James Onoja with Major Danjuma in tow, had cordoned off the Governor's Residence, where the Head of State was staying. Other officers, including Lt. Shelleng, and Lt. Dada (Paiko) took some soldiers from the group of about 30 on assignments, including mounting checkpoints to neutralize any incoming troops from Lagos, or Abeokuta (both less than 100 miles to the South of Ibadan).

However, Aguiyi-Ironsi, was not arrested until after 8 am. This may have been due to Danjuma trying to buy time, as he was essentially an 'outsider' now leading an operation in Ibadan.

He stalled attempts to blow up the building, perhaps believing that it was better to have the Head of State alive as a negotiation tool if necessary, but things escalated when the commander of the 2nd division (Ikeja Cantonment) in Lagos, Lt. Col. Njoku, who was with the Head of State earlier in the night, but was staying in an adjoining Guest House, shot his way past the Soldiers surrounding the building. Onoja, afraid that Njoku was going to get help to crush the coup, escaped, leaving Danjuma at the head of troops who barely knew him, with some suspecting that he was a Hausa speaking Igbo. He was about to completely lose the weak hold he had on the situation.

That Danjuma's Jukun homeland was in the 'North' was due to Lugard's fiscal and geographical gerrymandering, but delving into that would be tangential to the matters at hand.

Shelleng's return eased the tension a bit, and a survey of the grounds, close to the Guest House where Njoku had escaped from led to a conversation between Gowon, now Army Chief of Staff, and Danjuma. Gowon and Ogundipe (Brigadier, Chief of Staff Supreme Headquarters) had tried to reach Ironsi and Fajuyi without success, and called the Guest House). Gowon was able to confirm that Murtala (actually Lieutenants Pam and Paiko, but with Murtala's blessings) was ahead of him in Ibadan, and appealed to Danjuma that there should be no bloodshed.

Ironsi and Fajuyi then sent Ironsi's Northern Army ADC Lt. Sanni Bello outside to investigate the situation, and he (Bello) was promptly detained by the Coupists. Ironsi's Southern Air Force ADC, Lt. Andrew Nwankwo, remained inside. Fajuyi then steps outside to speak to Danjuma and the group, and is followed inside by Danjuma and about five fully armed soldiers.

killing of "brother-officers" in January and not bringing 'the so called dissident elements' to justice, and recounded how he

(Danjuma) had taken risks and run around to calm the ranks, but nothing had been done close to the end of July.

> *It is extremely important to note that the military officers (at least those ahead of the Belllo-Ribadu Officers) were not as concerned about the death of the civilian politicians in January. The killing of military officers, who were not involved in governance of the country, broke a brotherhood bond, and pointed more fingers back to Ironsi. The killings of the Brigadiers and Colonels who may have challenged Ironsi's decision to 'accept the invitation of the civilian cabinet' to lead, was a bitter pill to swallow for many of the officers of the Army. All of this played within a year of Ironsi gaining command of the Army as the GOC.*

Danjuma and the arresting team then marched Ironsi, Fajuyi and Nwankwo out to join the detained Sanni Bello, and attempted to take them to a secure location, but Paiko, and the Ibadan team resisted the move, as their own instructions from Murtala had nothing to do with Danjuma's 'no bloodshed arrangement' with Gowon.

Meanwhile, Lt. Col Murtala Muhammed was going round the military formations in Lagos, waking the conspirators up and telling them to get ready for action. He was almost pre-empting and sometimes countering Gowon, the Army Chief of Staff who was trying to rally troops to restore order and resist and prevent the killings. The relationship between Murtala Muhammed and Gowon was to take this form over the next decade and even beyond. Gowon kept trying to exert authority in a sometimes diplomatic way but remained quite conscious of Murtala Muhammed's popularity. However, Murtala Muhammed was always the subversive subordinate, having realised that he had 'street credibility' and could counter Gowon without consequence.

Murtala even wrote a letter to Gowon in the 70's, apologizing for all the times he had made life difficult for Gowon, despite his senior officer's effort to accommodate his excesses. Shortly afterwards, he (Murtala) sanctions or organizes the coup that removed Gowon from office as Head of State.

It is almost pointless listing out those involved in this mass murders of July 1966, as just about everyone you can imagine who's played a role in Nigeria since, including future Heads of State, participated in this orgy of killings. While the self-preservation argument has been pushed, especially once the killings in Abeokuta started, these killings were not in self-defence, as there was no clear and immediate threat to the lives of the perpetrators. More so, almost all the officers and men killed in the July 1966 Mass Murders had no business with the January 1966 Coup. The best some of the perpetrators of the murders could come up with was to accuse them of taunting. They claimed that Igbos were taking over, pointing to provocative songs and statements.

> *As has been pointed out and remains obvious, there is no real basis for the existence of a Nation for Nigeria and it is doubtful that one can be negotiated. The foreign created state has thus witnessed the entrenchment of systemic and routine cycles of violence and corruption that drags the indigenous peoples and the countries trapped therein to lower levels of poverty, illiteracy, extremism, and other clear forms of retrogression.*

There was action in all the other military formations and regional capitals, except maybe Benin City, although there were some initial signs of restraint and containment in Enugu and Kaduna. This would not last though.

Captain Martin Adamu, a key conspirator and loyal to Murtala soon took over the intelligence centre of the Army Headquarters. Brig. Ogundipe, Lt. Col. Gowon, Lt. Col Anwunah (Army Intelligence) and Maj. Mobolaji Johnson set up a rival decision-making operation centre at the Police headquarters, similar to the strategy of the British police Assistant IGP Marsden and High commissioner Francis Cuming-Bruce, who, six months earlier, had used the police headquarters and its Independent communication capability since the public phone lines had been cut.

From Left to Right sitting: Captain Adeyinka Adebayo,Captain Philip Effiong,Captain Imo,Major Samuel Adesoji Ademulegun,Major Wellington Bassey,General Forster, Major JTU Ironsi,Major Ralph Shodeinde,Captain Zakaria Maimalari, Captain Conrad Nwawo, Captain David Ejoor.

Middle Row standing: Lt Igboba,Lt George Kurubo, (non Nigerian standing next to Kurubo), Lt J Akaga,Lt Patrick Anwunah, Lt Louis Ogbonnia, Lt Chukwuemeka Odumegwu Ojukwu, Lt Eyo Ekpo, Lt Arthur Unegbe, Lt Abogo Largema

Back Row: Lt Hilary Njoku, 2nd Lt Nzefili, 2nd Lt Ogunenwe, 2nd Lt Shadrak, Lt Madiebo,2nd Lt Anthony Eze, Lt Yakubu Gowon, 2nd Lt Sylvanus Nwanjei,Lt Yakubu Pam, 2nd Lt Hassan Katsina.

Those not in the picture are Captain Kur Mohammed,Lt Victor Banjo and Lt Michael Okwechime

1st generation of indigenous Nigerian Army Officers[8]

[8] https://www.vanguardngr.com/2017/06/meet-first-30-nigerian-military-officers-before-1960/

Murtala Muhammed must have figured that there could be no coming back from the madness playing out at different formations all over the country. The plotters had appointed a certain Sergeant Dickson to secure the airport and commandeer a couple of planes that would fly the northern rebelling soldiers and their families out of Lagos to Kano once the carnage was done. No one could certainly get away with all the killings and carnage, Nigeria had been partitioned, and there was a speech ready for the announcement.

Ogundipe and Gowon rolled the dice and put out a quick reaction force to take control of the airport, but soldiers waylaid them at Ikeja. And in the shoot-out that ensued, about 30 soldiers were reportedly killed and sadly, some civilians, including some expatriates, such as the head of a major shoe company at the time. That was the end to any real opposition to the rebellion. It was now down to the inevitable partitioning, and maybe appeals and negotiation.

Nigeria was in a fluid state on Friday, July 29, and Saturday, July 30, 1966. The soldiers led by Murtala Mohammed had mostly achieved their killing and carnage objective in most parts of the country, and wanted to proceed with their partitioning of the country, but other than demands for everyone to return to their regions, they clearly had no clue how to get this done. They were the ones with the guns though, and in Nigeria, that has mostly been sufficient.

There were a couple of interesting (if that word could be used for such tragic events) bylines to the crazy killing orgies of this period. A second (compromised) quick response team was dispatched by Ogundipe and Gowon, under the command of a Second Lieutenant Usman or Osuma. He had probably joined the army courtesy of his Hausa-speaking abilities and due to the Bello Ribadu policy, but he was actually from the East; there were many others like him in the Army. 'Usman', like Onoja in Ibadan, skipped out of town, leaving all the craziness behind, with his life intact.

The military governor of the East, Lt. Col. Ojukwu, took preemptive action and left Enugu for Onitsha, probably because he was unsure of what was going on and maybe felt comfortable around his family business' operational transfer base. Or perhaps he wanted to mobilise personal security. After all, in the First Republic, Shehu Shagari admitted to having hundreds of thugs armed with weapons in Obalende and other parts of Lagos just in case a struggle ever broke out. However, they were not particularly useful to Shagari or the NPC when trouble did eventually break out in January 1966, similar to the legion of Fani-Kayode's thugs in Ibadan.

Ojukwu resurfaced in Enugu and mirrored the situation in Lagos by using the police headquarters as a base instead of the military barracks. This was nothing unusual anyway since he was a military administrator of the State and not the army commander. Ojukwu was also able to mobilise the Special Forces in Enugu as governor. Murtala Muhammed had contacted Lieutenant Yar'Adua to break into the armoury. But here, the commander Lt. Col. Ogunnewe had been able to take preemptive action that prevented the breakout of violence. Interestingly, the same approach had been attempted by Okonweze in Abeokuta but with very tragic consequences.

Ogunnewe managed to have a mixed group secure and guard the armoury and de-armed everyone else except himself and had all the officers at the officers' mess and all the soldiers on the parade ground, all watching each other in a chilling stalemate. This was a better solution than the Mexican Standoff Okonweze was said to propose in Abeokuta.

With the military checkmating itself in Enugu and Ojukwu in the safety of the Police Headquarters, He (Ojukwu) then started making demands of Lagos, especially Gowon and Ogundipe, who already had their hands very full with Murtala Mohammed, and the rampaging bulk of the rank and file of the Army.

In Kano, Lt. Col. Shuwa and Major Oluleye tried to maintain order to ensure that the ongoing chaos in Lagos and Ibadan did not spread there. The armory was secured before rampaging soldiers could get their hands on weapons. But some soldiers and civilians were killed anyway, and things got worse a few days later as civilians at the Kano airport trying to get out of town were killed by some NCOs, determined to join the orgy. Things soon got out of control, and Shuwa's attempt to get some Igbo Soldiers out of town ended badly with the execution of the soldiers by those meant to protect them. Shuwa soon had to escape from his own soldiers.

Shuwa, with Gowon.

Part of the reason for the anger of the Northern NCOs in Kano was rumors about Igbo Officers stockpiling private weapons. Searches by Shuwa and Oluleye, perhaps meant to dispel the rumors and diffuse tension, however found the rumoured private weapons.

Rather than this being proof of something sinister, the orgy of killings all over the country gave credence to the need for self-defense.

This captures the tragedy of the Nigerian State to a large extent. The indigenous countries and people, dealing with a forced and inefficient facade of unity in one conjoined State, spend more time worrying about defending themselves from attacks by other groups within and from outside the Nigerian State, than building prosperous and sustainable countries and societies. In many cases, the ability of these natural countries to exist, and perform is removed or suppressed.

In Kaduna, the Commander Lt. Col. Wellington Bassey was on leave, and his 2IC, Lt. Col Philip Effiong was out on Official Assignment. Lt. Col Alexander Madiebo, arrived from his Lagos base that morning, just making it out of the Lagos Airport before Murtala's killing posse seized the Airport. Sensing looming trouble, Madiebo was able to gather some Southern Lt. Cols and Majors to take preventive steps, so that when the killings started that night, a number of these officers were able to escape from Kaduna.

Many Officers were however still killed in gory manners by Bello-Ribadu Officers and NCOs. Some were however able to escape, due to the plans made earlier in the day, and the assistance from the Northern military Governor, Lt. Col. Hassan Katsina.

Major Obasanjo for instance, was reportedly smuggled out of town, in the boot of a car to Maiduguri. Lt. Col. Hassan Katsina provided Major Ogbemudia with a landrover jeep to facilitate his escape.

Major Ogbemudia had detained Lt. Dimka a few days earlier as he (Dimka) tried to break into the armoury. Perhap, Dimka was trying to be the one to set things off, to impress Lt. Col Murtala Muhammad, the Chief Plotter and Commander of Aure Paiko. Dimka was released to de-escalate the growing tension, but he had marked Ogbemudia for termination.

Upon learning of Ogbemudia skipping town, he organized a chasing gang of soldiers determined to kill the man that had detained him a few days earlier. Ogbemudia, refueled and may have stopped a bit at a Catering rest stop a few hundred miles outside Kaduna, unaware of this pursuit. This gave Dimka and the chasing party the opportunity to catch up, and the chasing and shooting reportedly continued all the way to Owo, over 500 kilometers away from Kaduna, where Ogbemudia, running out of fuel, abandoned the jeep, dashed into the bush, and was able to escape his determined hunters.

Just under a decade later, Lt. Col Buka Suka Dimka ironically and symbolically turned his guns on General Ramat Murtala Muhammad, the Head of the Nigerian State, with these words,

> *"Good morning fellow Nigerians, This is Lt. Col. B. Dimka of the Nigerian Army calling. I bring you good tidings. Murtala Muhammed's deficiency has been detected. His government is now overthrown by the young revolutionaries."*

Daily Times

Nigeria's most-wanted man

DIMKA

author of unholy Friday

(Turn to back page for story)

FOUND

Frontpage announcement of the capture of Lt. Col B.S Dimka after a Nationwide manhunt for the killing of Head of State, Gen. Murtala Muhammad. Less than a decade earlier, Dimka was the hunter in a literal cross-country manhunt, determined to kill, as part of a conspiracy to divide the country led by the then Lt. Col Murtala Muhammad.

Lt. Col. Ogunnewe's commendable efforts at maintaining the peace in Enugu were among the more impressive efforts along with those of Hassan Katsina and Shuwa at a time of complete chaos and killings.

Things settled down a bit in the next few days, and the negotiations and power play started. It was decided that non-Eastern soldiers will leave the East and Eastern soldiers and civilians in other regions be allowed to come back to the East.

Sadly, in another telltale sign of a young army with poor planning, strategy, and execution, it appears the same train was meant to make an Enugu-Kaduna-Lagos-Enugu loop. This meant Eastern and Northern Soldiers on the same train on sections of this loop.

Major Adekunle was meant to accompany the Northern soldiers from the East to Kaduna and the Igbo soldiers from Kaduna back to the East.

Yar'Adua was said to be in the center of some action where some Igbo soldiers were killed and thrown off the train. This may have happened due to Yar'Adua wanting to gain some "credibility" following the peace that was maintained in Enugu, which showed a failure to achieve his instructions to gain control of the armory.

Adekunle's effort to intervene almost cost him his life, but for the intervention of Capt. Gibson Jalo.

Ultimately, this was a poor operation as there should never have been direct interaction of soldiers from different regions. Different trains or multiple trips could have been used to prevent such an incident. Good planning and strategy can help overcome limited time and resources.

This was however, an Army that ultimately had no purpose, once its creators had left, and its objectives (British Interests and subjugation of the indigenous countries) had lapsed.

Nigeria's political leaders (the indigenous ones, especially) continue to follow the "Tafawa Balewa approach," which is a "balancing act of inefficiency," a haphazard approach of bring everybody in; have 80 ministers; development doesn't matter, corruption doesn't matter, just maintain some kind of balance and stability.

Yar'Adua's aggressive attempts to make up for his Enugu 'failure' in the following years did not impress some Bello-Ribadu officers, but only brought more resentment. They were the ones who earned and restored the dominance of 'The North' with the killings of July 1966, but Yar'Adua was the one, getting double promoted from Lt. Col to Brigadier, and becoming the Deputy Head of State following Murtala's death.

It was therefore not surprising that despite having the most powerful political machinery in the country (the Peoples Democratic Movement, which says it is not a political party), Babangida, and later Abacha (reported to have additional 'personal beef' with Yar'Adua) always managed to cook up excuses to ban him from the political process.

They may not have been able to prevent his rapid and privileged rise in the Military, but they were going to prevent him from becoming Civilian President of the Nigerian State.

Shehu Musa Yar'Adua's younger brother, Umaru, succeeded Obasanjo as Nigeria's civilian President.

Tragically, the Nigerian Nation has little to do with the constituent countries or people but remains a balancing act of corruption, inefficiency, foreign interests, and violent propagation. No real progress or sustainable societies can be forged out of this.

A cursory observer would notice that the events of January and July 1966 largely (not completely) seemed to be a confrontation between the East and the North, even if a lot of the action was in Lagos, Ibadan and Abeokuta, within the 'Afeire' Confederacy.

It is worth noting that the Western region was massively underrepresented in the Officer Corps, but especially in the NCO ranks of the Nigerian Army. This was why there was no thought or mention of any Northern soldier leaving Lagos, Ibadan, Abeokuta, or even Benin (in the recently created Mid-West), as soldiers were asked to return to their regions. The West was essentially "occupied territory", without its own soldiers.

This was somewhat cultural and also due to the strategic or historical role that this part of the country has played. A career in the Army was not regarded as a noble or desirable profession for the best and brightest from the region. It seemed like it was meant for those who were considered rascals and who, at least at the time, had no excellent career prospects. Interestingly, this was the strategy Benjamin Adekunle adopted, when asked to build the 43rd Marine Commando from the ground up, during the Nigeria-Biafra War. Hence the unique constitution of the Glover's Hausas and Lagos Constabulary, as described earlier in this book.

It is also a window into the cautious and somewhat cagey nature with which that entire part of Nigeria, referred to here as Afeire Confederacy[9] has always viewed and continues to view the Nigerian experiment. While it hasn't violently resisted it, it has probably been most slowed and negatively impacted by the Nigerian State. Many observers feel that once the Afeire makes up their mind that they are done with it, the Nigerian experiment will collapse.

They view their political figures who try to drag them into an English or Foduyan Fulbe-dominated centre with grave suspicion, whether it was Akintola or Abiola or Obasanjo or even as of 2021, Tinubu. Yet, they have always reacted angrily if they felt either of these leaders, except Akintola, was mistreated, having ticked all the boxes. Most of the southwestern part of the country has always been occupied territory in essence, as far as the military was concerned.

This, put in context, made Ojukwu's demands on Ogundipe and Gowon in the reality of 'Aure Paiko' as nothing more than tantrums and mere wishes.

The Igbo dominance of the Officer Corps, and the historical dominance of the NCO ranks by Hausas and Hausa speaking

[9] This can be taken to include the so-called Yoruba areas in Nigeria and 'Benin' Republic, and the related Igala, Edo, Esan and Itskeri Countries

soldiers from the proposed 'Central Region' pre-amalgamation (currently known as the middle belt), was the picture of the Army that NPC leader Ahmadu Bello, and his Defence Minister, Muhammadu Ribadu desperately tried to alter. Prime Minister Tafawa Balewa seemed relaxed in the fact that the man behind the rifle was Hausa, or Hausa speaking, and did not want to lower standards for the Officer Corps.

Another interesting side note from Lagos on July 29, was the young Captain Joseph Garba, who was very much part of the plots with Murtala Mohammed, but had with Paul Tarfa, heroically hidden Igbo colleagues at the National Guards Company (Brigade of Guards) in Ikoyi, on Lagos Island, as Abeokuta unexpectedly triggered 'Aure Paiko'.

Garba was called upon by Gowon and Ogundipe to try to quell the coup (that he was technically a part of), and was said to have made at least three trips between the Police headquarters on Lagos Island, where the loyal commanders were holed up, and the large 2nd Battalion in Ikeja (Maryland Cantonment) on the Mainland of Lagos where Murtala Muhammed, who was with, and had control of the bulk of the Army-held fort.

This would not be the last time that Joseph Garba would have to make runs back and forth and try to split the difference between Murtala Mohammed and Yakubu Gowon.

AURE PAIKO AND THE NIGERIAN STATE

Ibadan was suggested as an easy location to arrest the Head of State because it was majorly made up of a large contingent of Hausa Northern soldiers, who were seething with anger that three former Northern commanders of the local fourth division (Maimailari, Kur Muhammed, and Largema), were killed in the January 1966 coup. The fact that the plotters had not been court martialed or tried not only solidified the belief of Ironsi being the invisible hand behind Ifeajuna, but fueled rumours that the plotters could be released once Ironsi had significantly consolidated power.

There were riots in May 1966 following the unitary government declaration of Aguiyi-Ironsi that created 35 "provinces of regions." The Nigerian government today, though masked as a federal republic, is still effectively *that* unitary state. These riots were in Kano, Bauchi, Katsina, Zaria, Sokoto between May 4th and 5th, and they left about 600 Igbo civilians dead. That was a vital sign that the anger in the core Fulbe-controlled Hausa heartland was not just in the Army but also on the streets, spearheaded by the religious institutions and societal leaders.

It is interesting to note that this was only in the core Fulbe-controlled Hausa heartland. There were no riots in Borno, Ilorin, Makurdi, Igala land and other areas. This fact naturally indicates the different constructs of various parts of the gerrymandered North, a fact that official propaganda aggressively denied.

Aure Paiko eventually kicked off as Lieutenant Mwadkon Pam, or people close to him, ignored Murtala's order to delay the coup. The younger Pam was determined to avenge the killing of his older brother, Lt. Col James Pam, in January. Pam contacted Paiko (Garba Dada) and they got in touch with Murtala Muhammed, who would after deliberating a bit and realizing that things had already kicked off at Abeokuta, authorize the action in Ibadan.

It's never easy or accurate to draw stereotypes out of situations and events. Still, if one could create a stereotype of the Nigerian Army before independence, it would be one of the Igbo Officers commanding Hausa Soldiers.

When told by his own inspector-general of police that Easterners (technically ethnic Igbo Mid-Westerners, but this is a bit of Nigerian ITK) led most of the key strategic and combat positions in the Nigerian army, Prime Minister Tafawa Balewa replied that he was well aware of this, but that '*the man behind the rifle is a northerner.*' This was also why Muhammadu Ribadu and Ahmadu Bello made a case to the Prime Minister, shortly after independence, for the mass entrance of Northern soldiers to the officer cadre. The stereotype was certainly not imaginary.

It almost seems futile talking about heroes in such an orgy of bloodletting as *Aure Paiko* which was accompanied by the killings of tens of thousands of civilians afterwards.

It is however impossible to take a critical look at the first two coups d'etat and not commend Yakubu Gowon's role. He was indeed Army Chief of Staff when all hell broke loose, but he had distinguished himself at every point in that critical January to July 1966 period.

He firmly expressed the danger of Aguiyi-Ironsi taking power in January 1966. Even as the entire rank and file of the army were running rampant, he tried to counter, putting himself at significant risk, but maintaining the chain of command, deferring first to Aguiyi-Ironsi and subsequently to Ogundipe. It was also a considerable risk to counter the rampaging Murtala Muhammed who had pretty much partitioned the country since there was no coming back from the killings of killings of hundreds of soldiers and tens of thousands of civilians that he and the others had triggered.

Other officers who were worthy of commendation include David Ejoor, especially for his actions in both Lagos and Enugu in January, as well as Lieutenant Colonels Alexander Madiebo and Hassan Katsina, and Major Ogbemudia for their efforts at initially containing the spread of the orgy of killings to Kaduna. Be that as it may, they only had temporary success, and Hassan Katsina (with Major Abba Kyari) soon had to organize the escape of his fellow officers from Kaduna.

Lt. Col. Shuwa and Major Oluleye in Kano also deserve commendation, although their effort was only able to delay the bloodletting.

Lt. Col. Ogunnewe in the East, who maintained the previously mentioned mass stalemate.

Credit should be given to Lt. Col. Ojukwu as well. He withdrew from Enugu as the crisis started brewing, likely due to lack of information (you never know who is involved, and things would have been different if Yar'Adua gained access to the Armory), especially since he (Ojukwu) was a military governor and not necessarily an Army Commander at the time. His withdrawal to Onitsha, just 20 kilometers away from his

Nnewi hometown, may have been to quickly mobilize independent logistics and security resources that were available due to his father's wealth. He was able to copy the command structure that was trying to hold the fort in Lagos as he operated from the police headquarters, independent of any compromised army command structure.

However, Ojukwu soon became a source of distraction to the effort to manage Lagos, probably allowing his rivalry with Gowon to fester unnecessarily. He seemed to irritate those in Lagos who already had to deal with Murtala Muhammed, even as the latter had the backing of most of the army's rank and file, including the junior Bello-Ribadu officers who were now dominant in the second lieutenant and lieutenant ranks and had the immediate and direct command of the troops. So, Ojukwu's insistence on maintaining the army's order was not just an irritation; it was almost plain stupidity at that point. Some writers and biographers, including Frederick Forsyth often ignore these facts in their account of events. Ojukwu's effort to recruit Gowon, Ejoor, and Banjo into the 'broad based' Azikiwe plot is also often missing. This plot, which eventually failed to take off, probably created the chasm between Ojukwu and Azikiwe on one hand, and Ojukwu and Gowon, on the other.

Other commendable acts of bravery during these days of madness include the actions of Lieutenant Sani Bello, who was joint ADC to the Head of State alongside Lieutenant Andrew Nwankwo. Bello and Nwankwo were being driven to what seemed like their deaths, along with Ironsi and Fajuyi. Bello was however able to appeal to Dada (Paiko). They both hailed from the Minna-Kontagora axis on the southern fringes of Hausa land and the Northern area of Nupe country. Some other accounts suggested that Lt. Magoro caught up with the convoy, and had to plead for Bello's life. Bello then created a distraction, and based on a previously agreed signal, Nwankwo was able to escape into the bush. He then prevailed on the Soldiers not to go after him since he had achieved their objective in capturing Ironsi and Fajuyi.

Bello retired from the Army as a Colonel. As at 2021, he is one of Nigeria's wealthiest businessmen, an early investor in the largest telecommunications company in the country, the South Africa-owned MTN, and his son as of 2021, the governor of his own Niger state, which is primarily Nupe country and the previously mentioned Kotangora and Minna divisions of Hausa land.

As for Paiko, the young adjudant in Ibadan whose nickname (and hometown) became the codename for the bloodbath for which no one was held accountable, not much is known about his subsequent military service or record. Nigerian Military blogger BEEGEAGLES, in reporting his death in 2020[10], said he was a Senator representing Niger State in the 2nd Republic (1979 - 1983).

Danjuma, like Joe Garba, would be caught between Murtala Mohammed and Yakubu Gowon. By his own admission, he had tried to quell growing resentment in the Army, in the hope that Ironsi would bring the January 1966 plotters to book, but his presence at the Governor's residence to arrest Ironsi and Fajuyi probably took a significant part of the fight out of Gowon, as Danjuma and Walbe would have probably been Gowon's first option to checkmate any moves Murtala was making in Ibadan.

His closeness to both officers (but particularly Gowon) probably cost him the opportunity of becoming the Head of the Nigerian State after Murtala Muhammed's assassination. As

---

10 Another foremost Nigerian Military Historian, Dr. Nowa Omoigui suggests Rtd Col. Garba Dada (Paiko) died much earlier, at some point before 2003.

such, a compromise candidate was reached in the person of the highly reluctant, and some say terrified, Olusegun Obasanjo.

As the impact of Murtala Muhammed's decision to partition the country[11] and *Aure Paiko* began to unravel, the English represented by the British High Commissioner, the Super Perm Secs (as the core of about 3 to 5 permanent secretaries and civilian administrators were known), and Gowon took the central stage to reverse the partitioning of the Nigerian experiment.

The Perm Secs, the English ambassador, and the American ambassador were able to convince Murtala to reverse the already successful succession and partitioning, with the condition that the more stable Gowon became Head of State. Gowon, who had been arrested by Murtala and his mutineers upon his arrival in Ikeja, emerged as Head of the Nigerian State,

It should be reiterated that it wasn't up to this 'kangaroo council' (describing it as anything but, would be a travesty) to determine the fate of naturally existing countries.

Murtala shows up in Mufti (Civilian Clothing) to a Press Conference by the newly minted Major General Yakubu Gowon in 1968

11 Some refer to 'Aure Paiko' as Operation Araba. Araba is the Hausa word for separation or partition, and this was the objective of the plotters.

The English would always act in their interest and so really cannot be faulted (but must be viewed with extreme caution), after all, everyone has to do proper analysis and work for the best interest of their people. From the English perspective, Nigeria had moved from just another pawn in its competition with its European friends in the scramble for Africa to a place to get soldiers to counter Germany in the first and second world wars. After the second world war, Hausa country, especially, became the most important piece to guarantee its insurance in the case of a third world war.

In the first military coup, England danced around the request for military assistance from Dipcharima, waiting to support whoever emerged victorious from the power play. Then, Nigeria was still important enough to be its fourth-largest embassy globally, and it even considered sending the Royal Prince down to Nigeria to be its ambassador. The events of 1966 made England step back a bit. Then, Araba/Aure Paiko made it seemed like power was swinging back to a Fulbe hegemony that had always danced to its (English or British, if you prefer) tune as a fellow foreign occupying power in the indigenous country of Hausas and, by extension Nigeria. The timing was key, as Crude Oil was beginning to play a more important role in global politics and Economy. So, the English were only acting in their interests in keeping Nigeria together and supporting the North(ern Fuduyan Foreign Fulbe Hegemony).

GOWON, OJUKWU, MURTALA AND AWOLOWO

As previously mentioned, the central narrative, even if unofficial, in the Nigeria of the late 1970's and 1980's was to portray Gowon as the evil man that the villain Dimka was trying to bring back to power after the killing of Nigeria's saintly hero, Murtala Muhammed. The reality, however, doesn't match that narrative and perception. There was little in Gowon's career or character that suggested that he would have trusted the very unstable Dimka with such a task, even if he was behind a plot to unseat the Murtala Government.

Ojukwu, having to deal with the slaughter of hundreds of Igbo Officers and Soldiers and tens of thousands of Igbo civilians in various parts of the country, found Gowon's emergence as the Head of State unacceptable. It didn't help that he was also removed from Murtala Muhammad and the events in other parts of the Nigerian State.

Every statement or broadcast from Gowon seemed to be met by a counter statement or broadcast from Ojukwu. The bantering continued even up to counterclaims of each taking credit for the release of Obafemi Awolowo from the Calabar prisons.

In every way, Yakubu Gowon was a more moderate and restrained option than the Murtala of 1967, which makes it disappointing that an avoidable Civil War was not prevented.

A closer look at these four gentlemen in this crucial moment of Nigerian History is quite revealing.

Murtala, whose father was from Auchi, but landed in Kano via Agege, Lagos, was very much the arrowhead or Swordsman of the foreign Fulbe controlled Hausaland. Ahmadu Bello promised to return to Lagos with a Sword. That promise was fulfilled by Murtala in the last days of July 1966.

He was very much the Military might in this situation, and represented the mindset of "domestic" colonials that believes it owns Nigeria. This group has no real interest in Nigeria, outside of control and plundering. It does not believe it is subject to the laws, and has no real interest in sacrificing for Nation-building. At this point, its objective was to get out of Nigeria as quickly as possible, as retaining control of Hausaland was better than an Igbo domination of a Unitary Nigeria.

Going back to 1966, the English Government, the senior civil servants (who can be viewed in some sense as successor to the Colonials) and Yakubu Gowon convinced Murtala and the Araba group to change its mind, since control over the 'subjugated south' as part of the whole country, was better than the 'Araba' that 'Aure Paiko' was meant to achieve.

Under the Buhari Administration from 2015 (or because of it, given Buhari's outlook and 'perspective') the Fuduyan Fulbes have actively collaborated with the (Nigerian) Kanuris in the control of the Nigerian State, following a century plus war in which the Fuduyan Fulbes failed to gain control of the religious and traditional institutions in Kanuri Country.

It must be pointed out that Murtala Muhammad showed signs of being a different person during his brief time as Head of the Nigerian State. He helped formulate a determined "Africa first," anti-Aparthied, non-aligned foreign policy for Nigeria, which was continued under Obasanjo, and saw Nigeria take aggressive steps to show its seriousness and commitment to this policy. This included the nationalization of British Petroleum, follow the company's defiance of the instruction not to load Crude Oil from the Niger-Delta for the Aparthied South African Market and Government.

Maybe it was Murtala maturing, maybe it was just the initial 'gra-gra' of getting into office, maybe it was the rough reminder about his identity, following the negotiations to sanction him as Head of State following the overthrow of Gowon by the Bello-Ribadu crew. He even made a trip to Auchi, and became particularly close to Ibrahim Kagara (Taiwo) for reasons that seem pretty obvious. Murtala seemed to be a changed man in 1975

For the first seven months of 1966, **Lt. Col Yakubu Gowon** showed what is possibly the finest form of soldiering and military leadership that the Nigerian State had ever witnessed till that point, and possibly since. Gowon had been sterling, excellent and ultra professional, often swimming against the tide, occasionally walking on eggshells, trying to maintain what was sometimes a delicate balance, and at other times a raging inferno that could quickly turn around and consume him as well.

Gowon meets Awolowo upon his release from Prison, with Murtala watching closely

His actions as a Statesman did not meet the very high standards that he had set as a soldier over the previous 6 months; and he played a significant role as the English (represented by Francis Cumming-Bruce) and the Super Perm Secs' worked to reverse Murtala Muhammed's partitioning.

Gowon is Angas, whose home Country stretches from the Jos Plateau area to the current Federal Capital Territory in Abuja. Even though the Angas are indigenes of Jos, they are a minority of minorities, if that term can be used. When you think of Jos and Plateau, you think of the Beroms and the Taroks. And when you think of the FCT, you think of the Gbayis (or Gbagis) and the Nupes.

Gowon had to manage the Bello-Ribadu lieutenants and second lieutenants running amok with the largely Hausa-dominated (but with a significant number of middle-belt) NCOs in most of the Army formations in the previous days. His only shot at any relevance and probably survival lay in the united Nigerian state. In an Arewa Republic, or the Republic of Northern Nigeria, his position as a minority of minorities would be more than evident.

If he could barely contain Murtala Muhammed in the context of a political and military Nigeria, he would have no shot in the Republic of Northern Nigeria. As such, **G**o **O**n **W**ith **O**ne **N**igeria had as much to do with Gowon's survival and relevance as it did with keeping the English and Fuduyan Fulbe contraption together.

Thus, Gowon the soldier was excellent, Gowon the Statesman had his own biases, and some of these were self-preservatory. Given his ultra-minority profile, there was no way he was going to be a part of a Northern Arewa Republic with (or under the control of) Murtala Mohammed and the "out of control" Bello-Ribadu Lts and 2nd Lts.

From a military perspective, Western Nigeria was already occupied territory, given the history and development of the Nigerian Military (Glover's Hausas) up till that time. The region, fairly defined as a confederation of sub-ethnicities that were just rounding up their own internal civil wars (Ibadan had just negotiated a settlement with related Ekiti and Ijesha in the Kiriji war about the time the English Colonials were consolidating) and seemed ambivalent about a Nigerian State that had left almost a quarter of the country outside its borders.

The Action Group crisis, and the location of the 3 major military divisions (Lagos, Abẹokuta and Ibadan) meant there was likely to be no problem from this region.

The Eastern region, even if not a subdued country, had just just lost significant military manpower, was definitely scarred by all the violence and killings that its people had suffered, and the capacity to push back militarily at that time, especially in terms of personnel and to a large extent morale.

So Nigeria made sense for Gowon, especially since he was now the negotiated choice to become the Head of State.

But, perhaps, his rational fear of the Arewa republic could have been placated with the middle-belt or central region within a Confederation of Nigerian States that gave each region the opportunity to live life and develop as it chose, but maintained the minimum terms of a Confederation agreement. Perhaps he had not seen the map on the cover of this book which predates Lugard's amalgamation or even his replacing one Atahiru with another Atahiru in Sokoto. The map was a standard consideration for the Colonial Office, but was not the only one. Suggested territories and regions ranged from four upwards as realistic and individual subdivisions of a Nigerian Confederation that was definitely going to be Africa's most complex post-colonial state.

Obafemi Awolowo initially refused to join the fray, wanting to play the fatherly role of reconciling the young colonels and averting further disaster in the country. He even took a famous trip to Ojukwu in the East to try to mediate the crisis. However, as recounted in a couple of accounts, including Wole Soyinka's 'You Must Set Forth at Dawn', Ojukwu either reneged on an agreement or moved ahead of a deal that he should maintain the status quo for a couple of weeks (this was after the Aburi Accord) and not declare an independent republic.

Ojukwu seemingly countered this by suggesting that his last words to Awolowo was 'Baba, Atilo', meaning "Old Man, we have left."

Awolowo was under pressure as a leader without troops. He had also previously turned down Gowon's request to join the government, even when Gowon asked Mrs Awolowo to convince her husband[12]. However, he would eventually join the government of General Yakubu Gowon in what was the most

---

12 https://guardian.ng/news/awos-wife-convinced-him-to-join-my-government-says-gowon/

significant blunder of his entire public career. It was excellent for the Nigerian State, but not for the sovereignty of the indigenous peoples and countries, which is what Awolowo stood for, and resisted the British and Ahmadu Bello for in the 1950's. Over the next couple of years, he helped make the Nigerian State the strongest and most secure it had ever been.

A negotiated confederation would have been the best option for the Nigerian State at this critical time, and ever since. That was what the four main actors wanted, without stating it, and for some, possibly without knowing it. Apologies for the brutally honest analysis (as you should be used to on these pages by now). You won't resolve a problem by a dishonest analysis of the issues.

The dispute between these very young men, not necessarily Aburi, eventually led to the civil war; everyone acted in their own interests. It was in Gowon's interest for One Nigeria to continue, as it was in the interest of the English. At this time, given what had occurred, it was probably in the interest of Ojukwu to form the Republic of Biafra even though the Ndigbo sometimes say that they are the greatest proponents of a united Nigeria. It was certainly in the interest of Foduyan Fulbes (not Fulanis all over West Africa, but certainly the Foduyan Fulbes who ruled Hausa land) for Nigeria to continue existing. A unitary or even federal Nigerian state was not in the interest of Awolowo or his people.

While Awolowo may have lacked military muscle, it would probably would have been better if he had yielded the position of deputy-whatever-he-was under Gowon to Enahoro or someone in his camp, even though he brought a few of them into the government. But it was him that Gowon wanted for credibility reasons. He performed credibly well in the position; the war was financed without any debt and with relative fiscal discipline. More so, he left as soon as the war was over in 1971. Yet, his occupying the position probably was not in his own interest or the interest of the people who now perceived him as a willing participant in that war when his part of the country was pretty much-occupied territory.

If there was anyone standing up for the victims of the mass killings of July and August 1966, Igbos and other peoples who were attacked as 'Easterners', it was **Lt. Col. Chwewuemeka Odumegwu Ojukwu.** He had definitely heard about the horrific ways in which officers and men from that region were killed. Some, locked up in a guardroom, possibly for their own protection, had grenades tossed into the room and blown to bits by those determined to kill them. The mass killings of soldiers soon spread to the streets in several cities as thousands of civilians were killed.

No group deserves to go through all of this, largely due to the actions, and inaction of Major General Ironsi, as Head of State and for taking over power in the first place, and the fact that there was no talk or possibility of bringing the very identifiable perpetrators of these horrendous crimes to justice was more than enough ground for any region to seek to leave the fraudulently created Nigerian State.

The Ndigbo embraced the emerging Nigerian State and the opportunities it provided, and dominated the mid-cadre of the Officer Corps of the military, at the time of the Independence. Colonialism also opened up the previously insular, but very republican Igbo country, and the people, used to competition for ranking privileges within age groups and societies, poured out to take advantage of new frontiers.

The first military coup happened within a year of Ironsi at the helm of the Nigerian Military. Ironsi, who had been ranked 3rd (among 3 candidates) for the job by the departing GOC was appointed to the position on the insistence of President Azikiwe, as he and Prime Minister Balewa tried to resolve the several issues they were bickering over.

As such, the outcome of the first coup, opened the door for a boatload of questions, especially as two of Nigeria's (four) indigenous generals (and another full Colonel) were dead, with a third in London as the Defence Attache. There was thus no ranking military voice to oppose Ironsi's takeover, or propose options for the Nigerian State at this time.

The delay or failure to try the coupists led to growing dissension within the Military, and the Unitary tendencies of the Ironsi regime (including the interference in the Northern region civil service requirements, which Lt. Col Ojukwu suggested would provide more opportunities for 'Easterners') increased the anxiety of 'Northern' opinion leaders.

This was how many NCOs and officers from the middle-belt areas, with homelands experiencing of the expansionist tendencies of the Fuduyan Fulbe rulers of Hausaland (TIVs for instance), participated in 'Aure Paiko', but were not keen to be part of Murtala's conclusion to Operation Araba.

This in no way attempts to blame the victims of 'Auta Paiko', but it is useful to provide the proper context for these tragic events.

Considering that there was no central purpose or underlying reason for the creation of Nigeria other than the plundering of the indigenous countries by foreigners, the demand for Biafra was certainly reasonable.

Yakubu Gowon and Emeka Ojukwu in Aburi, Ghana

Too much consequence is often placed on Aburi. Gowon was in a position of strength as far as Nigeria and Biafra was concerned, but was in a position of great weakness as far as Nigeria and Arewa (operation Araba was concerned). You always have to consider your opponent's strategic choices, but Ojukwu seemed to show little sensitivity to the fact that Gowon had to deal with Murtala, the Bello-Ribadu Officers, and the bulk of the Army NCOs.

The Aburi Accord (Decree 8) seemed to focus too much on what was meant to be an interim Military arrangement (Military Administrators and the Head of State), but the meeting was a start.

Ojukwu's failure to attend the ratification in Benin City where they could have ironed out differences (subsections 70 and 71, which gave Gowon controlling rights) of the situation, and agreed on something definitely much better than a bloody 30 month killing spree.

The situation before these young men was not one of course-mates seeking equal share. It was the lives and well-being of millions. Ojukwu believed himself to be superior to Gowon with a very good basis. He was the Oxford educated officer, who had joined the Army after a couple of years working as a civilian. His joining the Army as an NCO, was to rebel against what he felt was his millionaire father's influence over his life, so in terms of experience, and the rank and commission he felt he would have had, if he made the decision to join the Military earlier, he was Gowon's senior. It didn't help that Gowon, who Ojukwu viewed as inferior, had now been negotiated into position as Head of the Nigerian State. This was not a Military Commission, it was a quick arrangement by the creators and owners of Nigeria. Murtala Muhammad as the swordsman of the Fuduyan Fulbes (also known as 'Northern Nigeria', but with the recently created the term 'Hausa-Fulani' now adopted), and Francis Cumming-Bruce, representing the English Colonials, who brought Albert Mathews with him. Mathews was Ambassador of the United States, the global superpower to which the English were now subservient, but who generally viewed African matters in deference to England and France or from an Anti-Communist or Anti-Socialist perspective.

The civil servants were also there to take care of the paperwork, after the fact.

As such, despite the extremely tragic and upsetting circumstances, Ojukwu needed to view the situation, and Gowon, for what it was, and not keep making counter broadcasts and demands that would probably not be met. He had accepted an extra-Military arrangement as Governor under Ironsi. He could not publicly turn around and start insisting on Military ranking and seniority in this extraordinary situation, except he was demanding the end of the Nigerian State, and prepared to follow through with it, and take on everyone (and Nation) that had negotiated the Gowon "solution."

A lot more could have been done privately though, and Benin was an opportunity for that. David Ejoor, who was certainly not subservient to Gowon, was hosting.

Ejoor and Banjo in training as Cadets. Banjo led the Biafran Army in its most significant advancement into Benin, where Ejoor was military Governor. Banjo's concern about the capacity and discipline of his troops to execute an advance on Lagos, reportedly led him to attempt an unauthorized negotiated solution to the war, again from Benin. He was executed shortly afterwards.

Gowon the statesman would have been partly driven by self-preservation instinct, like just about everyone. He was inexperienced and limited in political matters, but he was sincere, brave and conciliatory. The calibre of people and forces settling on him as leader of the Nigerian State also meant that Murtala had been told to stand down, and let Gowon operate.

It was now time to reach out to his old friend and coursemate, and try to work things out. Positions and interests were changing very quickly in these times. The middle-belt officers for instance, despite clear anger and opposition to Ahmadu Bello's incursions to Tivland, had participated in 'Aure Paiko'. One of them, Lt. Pam, eager to avenge his brother's death, triggered the whole thing off. They however had no interest in the 'Operation Araba' conclusion of the process. That was for the Fuduyan Fulbe's, their Hausa subjects, and maybe the Kanuris.

Gowon and Ojukwu: The young men who needed to find a way to prevent carnage.

Gowon just needed time, and perhaps privacy, with his old friend, who, by the way had invited him and Banjo to be part of a plot supposedly linked to the President, a couple of years back. It was in this spirit that Gowon went to Aburi. Ojukwu was however prepared. Gowon had clearly deviated from the objective of those who put him in office, with the Aburi (Accord) agreement, so a slightly different variant was released.

Ojukwu and Gowon: Those young men were gone, and all that was left, were these old men (© Red)

Ojukwu's press conferences, counter conferences and insults on Gowon continued. Some have jokingly suggested that Gowon's enrolment in a doctorate following his removal as Head of the Nigerian State, flowed from deep pain over Ojukwu's insults.

It seemed Ojukwu did not realize who had the advantage, in terms of military might, external superpower & past power advantage etc. Benin City, not Aburi was the last opportunity to avoid even more tragedy and carnage, and one man, who was completely outside the process officially, had the credibility and influence to set these young men, who were now bickering over everything, including who should take the credit for releasing him from prison, on the right path.

Obafemi Awolowo had been through a very rough 12 to 15 years. He had been seemingly harassed, double-crossed, betrayed, stabbed in the back and eventually jailed by anyone and everyone that seemed to matter in the Nigerian State, and its colonial predecessor.

He spent most of the 50s trying to resist the alliance that was being instigated by the English Colonials with the Fulbe rulers of Hausaland represented by Ahmadu Bello's NPC, preferring to build his own political base by collapsing areas and divisions like Offa, Kabba and Ilorin into their natural home country, and building alliances with Joseph Tarka and the UMBC.

However, by similar logic of ethnic nationalities, facts of history and geography, the Western region and Awolowo were also vulnerable to threats of natural partition and division within its region. Many divisions on the eastern front of the region, including the areas of Asaba, Ibusa, Ogwashi-Uku formed the Anioma and Ika areas of Igbo (speaking) country[13]. It should be pointed out that several notable advocates of the 'Igbo-speaking' school of thought, including Philip Asiodu, now suggest that these groups, especially the Anioma, are Igbo, and not just Igbo speaking[14].

As such, it was not surprising that the peoples in these communities wanted out of the Western region, as did the Western Ijaw and Urhobo communities and ethnicities, with Dennis Osadebey (Anioma-Igbo) and Samuel Mariere (Urhobo), becoming the Premier of the Mid-Western region. The reality of Colonialism, and the regions it created, and indeed the Nigerian State itself, was that previously unaffiliated communities were lumped into legal, fiscal and geographical entities.

The Action Group recognized the desire and agitation for a Mid-West region, and had moved a motion in 1955 for the creation of **Ben**in and **Del**ta[15]. It was hoping that it would make

13 There has been some decades-long controversy over whether these areas (including the Ikwere areas of Rivers) are Igbo or just Igbo speaking.

14 https://www.vanguardngr.com/2020/05/anioma-v-igbo/

15 The name BenDel state from Benin and Delta provinces was adopted in 1966. Babangida in 1991, separated Bendel into 2 states, and included Warri in Delta State, while choosing Asaba, the hometown of his wife

the move, shortly after independence, as the party in control of the country in a likely alliance with the NCNC.

Awolowo did not oppose the move, but was not in favour of the Benin province, Akoko-Edo and Warri being part of the Mid-West region.

The move to restore the Olu of Warri title, as it was reportedly previously known and likely influenced by the Ologbosere (traditional Prime Minister) Ogbemi Rewane, only seemed to infuriate Ijaw and Urhobo communities in the town and area.

Anthony Enahoro, Alfred Rewane (younger brother of the Ologbosere of Warri), Obafemi Awolowo and Arthur Prest

Awolowo had emphasized utmost respect and preservation of the cultural institutions, even as society quickly needed to modernize and develop, soon had to placate the Oba Akenzua of Benin (Omo n'Oba N'Edo) who was showing strong NCNC leanings and support for the Mid-West State with Benin City as the Capital. Awolowo had already deposed the Alaafin and lost

Maryam, as the capital of the State. Benin City remained the Capital of the new Edo State.

a key ally to a clash with a traditional ruler[16]. Taking on the Oba of Benin (who alongside the Alaafin & the Ooni could be regarded as the Cardinal Monarchs of a region with about 30 other ranking Monarchs) would be unwise, and on a personal level, the Midwest region would would literally separate him from two of his closest allies, Anthony Enahoro and Aurthur Prest.

Eventually several aligned interests (Akintola battling for survival, NPC and NCNC support and strong support of the citizenry, with over 95% of the votes supporting the plebiscite) led to the creation of the Mid-West region, and Awolowo and Enahoro ended up in jail, along with most of the rest of the AG leadership.

The Action Group had been universally acknowledged for proving that an excellence in African Governance, that quickly translates to an improvement in the lives of its people, while preserving or even enhancing the role of indigenous cultural and traditional institutions (admittedly, having much deeper significance in its region than elsewhere in the Nigerian State) was possible. Despite general ambivalence towards the Nigerian State, it believed it had a path to the leadership of that State. This included a consolidation of its core base, reeling in the parts of its base that were gerrymandered into other regions, and getting a coalition on its own terms (not on Foreign English Colonial or Fuduyan Fulbe terms) with the NCNC.

With Dr. Azikiwe's volte-face and preference to be junior partner to the NPC, and internal dissention from Akintola and Fani-Kayode, the party, and its leaders were easy pickings for several opposing parties with no common interest, other than enmity for the Action Group.

---

16 This cannot be proven, but 'everyone' believed the Alaafin diabolically killed Bode Thomas, and the Monarch did little to correct the impression.

But here he was in late 1966, out of prison, and with enough credibility to bring the bickering young men together. Could he succeed where even the likes of Haile Selasse and other African leaders had failed with Ojukwu and Gowon?

While it is unlikely that Awolowo would have wanted to be involved with an English led (or influenced) intervention, it is also unlikely that Gowon would have invited Awolowo to his Government without notifying the English, at the very least.

Ojukwu missing the Benin meeting, sharp differences over what was agreed in the meeting between Ojukwu and Awolowo (an agreement not to secede, or not to announce secession yet) was claimed, and Gowon getting Mrs. Awolowo to put pressure on her husband eventually pushed Awolowo from an interventionist role to Finance Commissioner and unappointed strategist for the Nigerian side.

Benin City would have given Ojukwu, Gowon, Ejoor and Banjo to sort out whatever issues they had between and among them. Victor Banjo had again bizarrely offered or accepted to 'help out' in a government where he clearly was an outsider, and Ojukwu could have brought him to Benin for some recognized muscle on his own side of the negotiation table. Hassan-Katsina and Adeyinka Adebayo, were the other Military Governors at the Benin meeting.

Ojukwu's ignoring of the meeting of Gowon and the regional Governors in Benin, meant a hardening of stances across board, and Gowon's decision to create 12 States by military fiat/decree, was the first of many by 'Northern' Military officers which has taken Nigeria to the current (as at 2021) 36 states, which essentially function as branches of the Central Government,
with the Constitution crafted by the military severely concentrating power in the Center through an overreaching "exclusive" list.

Gowon's move in early May 1967 was clearly in bad faith, and an attempt to whittle Ojukwu's sphere of influence and create alternative administrative structures. Ojukwu responded towards the end of the month with the declaration of Biafra.

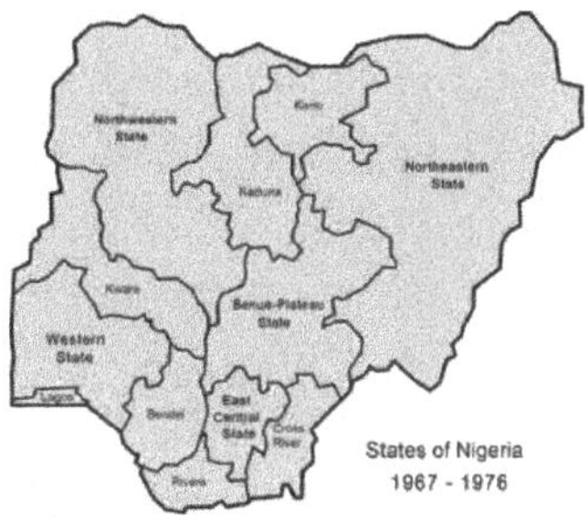

Awolowo did not go into the Gowon government alone. His allies from the first republic Joseph Tarka, and Anthony Enahoro (also released from jail) held cabinet positions as well. The thought that Awolowo could have recommended either one of these fine gentlemen and stayed outside the fray, lingers.

Enahoro, with Awolowo in the service of Gowon's Government.

With a knack for efficient execution in governance, Awolowo quickly put the Nigerian State in what may be the strongest fiscal position it has ever been throughout its existence. A decade earlier, he needed to raise taxes to finance Universal Primary Education and Child Healthcare. This time, Oil and a bit of strategy made it so much easier.

Four factors can be highlighted:

1) Get Revenues: Ojukwu wanted £2 million in monthly royalties from Shell BP. He had threatened sanctions, but couldn't back it up as the companies offered an eighth of the amount demanded (£250,000). Awolowo was able to get the companies to pay all the royalties to Gowon's government, which then enforced a sea blockade, and eventually quickly captured the oil producing Niger-Delta.
2) Efficiently allocate expenses to the areas of need. Divisional Commanders were responsible for arms procurement for instance. This ensured there were no complaints about HQ procurements, and certainly limited the types and volume of leakages. There were probably still leakages, and Akinriade suggested the divisional commanders were unnecessarily dragging things on, which made the mid-rank officers reach out to brothers across the line to end hostilities. Compare this to the situation (in 2017) where undelivered arms contracts worth a quarter of a billion dollars are awarded to the right type of 'migrant or stagnant' foreigner. A Nigerien this time.[17]

---

[17] https://www.news24.com/news24/africa/news/us-court-battle-gives-clues-to-nigerian-arms-scandal-20170518

3) Accountability and Savings: The spend it all mentality of future Governments was absent. With Awolowo at the helm, the Government reportedly saved £990,000 that was due monthly to the newly created East Central State (much higher than the Oil companies offered), and handed it over at the end of the War.
4) Delegation and Support: Awolowo also had the support of the 'Super Perm Secs', including Asiodu and Arikpo who fell in line behind the experienced Administrator. The Federal side was thus unexpectedly efficient, while Biafra seemingly tended towards dictatorship.

Despite Awolowo's achievements for the Nigerian State, there are genuine questions regarding the fact that he help create, or stabilize, the Central/Federal Government as the "monster" that is the primary revenue collection agency, and then distributes to its branches (that it arbitrarily created) as it deems fit.

His defenders will argue that the 12 states were created before he became part of Gowon's government, but Awolowo had the big picture, and needed to find a way to enforce it, despite the relative inexperience of Ojukwu and Gowon. Murtala soon bungled his way out of relevance (going abroad for a short while) following a couple of spectacular war errors, flowing from his stubbornness and not realizing that conventional warfare was different from catching brother officers and soldiers by surprise in their homes and barracks.

The fiscal and revenue generating arrangements that were put in place, was for a Nation at war, and countering a seceding component. The process to reverse this ad-hoc arrangement, which remains till date (2021) should have been put in place, rather than perpetuate the monster that encourages laziness and corruption, and discourages productivity and sustainability for the countries trapped in the Nigerian State. Sustainability was hardly a factor for the States (branches of the Central Government) created by military fiat as well.

Awolowo was also able to provide grants to the African Continental Bank, and the Cooperative Bank of Eastern Nigeria to help small business owners kick things off, after the war. The change in currency, like securing the oil revenues at the start of the war, proved critical in limiting the ability of the Biafran State. Ojukwu would probably have wanted an opportunity to undo, or unsay whatever it was that pushed the hardly temperamental, but straight shooting and straight talking Awolowo, over the line from mediation to working with Gowon.

Awolowo's service to the Nigerian State, also came with personal image, and reputational blows, as his response to Biafran soldiers diverting food aid meant for the populace was to suggest a complete shutdown of the food aid program to squeeze the Biafran Army, since the food reportedly wasn't getting to the populace. No amount of explanation or pleas for forgiveness can excuse the suggestion and execution of starvation as a war policy, even if it is said to have helped end a war that should have been prevented at all costs.

The exchange of Biafran currency, whose exchange value, basis for printing and volume in circulation, could not be determined, was also problematic, but it is not uncommon for fiat money issued by a troubled state to become worthless. Hopefully, this stark history will not be repeated for citizens of Nigeria, as the naira is is something of a freefall in 2021, with the top treasury official adding petrol to the fire by removing transparency from what is the 'real market' for everyone not benefitting from the official 'arbitrage, round tripping' system.

It could not be verified if the old Nigerian currencies were traded in, as there would be no excuse for that not to have been done. The fact that the change of currency severely affected the Biafran State also raises questions about the level of adoption of the Biafran Pound.

Awolowo stepped down from Gowon's government in 1971, and he was replaced by Fuduyan Fulbe key player, Shehu Shagari, who had joined the cabinet in an economic development role, after the civil war in 1970. As expected, fiscal discipline and savings soon gave way to Cement Armadas[18], with unbelievable demurrage terms, and general wastage. Despite increased oil earnings, from around 1973, the nation had major problems financing the Delta steel company, Aladja in 1979[19] The Shagari Presidency cannot be in the conversation, given all the economic troubles.

Serving in the cabinet with close pre-independence allies should have been an opportunity to jointly prepare and strategize for the political process, whenever it returned, if it ever did. He had his President and Vice President right there.

Instead in 1979, neither Enahoro nor Tarka, even bothered joining forces with Awolowo's UPN, prefering to go with Shagari's NPN. Aper Aku emerged as Governor of Tarka's home State of Benue, with his blessing,on the banner of NPN, while Ambrose Folorunso Alli became Governor of Enahoro's Bendel State on the banner of UPN, presumably without Enahoro's support.

Murtala Mohammed, as military dictator, further splintered Nigeria into 19 States.

The only position for Enahoro in Awolowo's UPN would have been that of the Presidential candidate, with Awolowo

---

18 https://www.history.org.uk/historian/resource/10169/legacies-of-the-cement-armada

19 https://www.euromoney.com/article/b1f5w54yw0gxzh/the-extraordinary-nigerian-jumbo

taking a role on the sidelines in a party building and/or advisory capacity.

Instead, the 70 year old Awolowo, and the 75 year old Azikiwe, were trying to create an alliance, two decades after the fact. While younger comrades and jailed 'Action Group' members like Lateef Jakande and Victor Onabanjo became Governors of Lagos and Ogun State, Awolowo's run for President in 1979 raised questions about his legacy.

Azikiwe, with Awolowo, met regarding the 1979 elections.

There were no questions about his motives. He wanted to do good, and knew how to deliver. Most times, there were no questions about his methods as well. They were practical, straightforward, and efficient. At 70, did he have to do the 'good' by himself?

He had been praised for focusing on his region and generally avoiding the obvious mess that Nigeria was going to be, in the 50's. The English Colonials were shocked by his sudden National outlook and choice to go to the Center, as opposition leader in the first republic, to 'facilitate a healthy democracy', according to him. Maybe he always had Nigeria in his sights, and the move to the centre was to await the collapse of the very poorly performing, and struggling coalition, and be right there to pick up the pieces when the eventually happened.

Ifeajuna and company happened instead.

POST CIVIL WAR (MILITARY) HISTORY

1970 - Gowon made a 'No Victor, No Vanquished' declaration regarding the war, and launched a 'Rehabilitation, Reconstruction and Reconciliation' program for the country. A four-year development plan is also launched.

General Yakubu Gowon

1971 - Awolowo resigns as Finance Minister and unofficial Vice President. Gowon said he (Awolowo) had helped Nigeria get out of financial disaster, and finance the war from internal resources, without debt, or devaluation of the currency.

Enahoro and Tarka also served in Gowon's wartime Cabinet. Awolowo is replaced as Finance Minister by Shehu Shagari.

1974 - Gowon postpones the planned handover date to civilian rule from 1976 to an unstated time in the future. Student protests in University Campuses led to closure of Universities.

1975 - The Bello-Ribadu Officers led by Yar'Adua decided that it was time to take over 'their' country and approached Gowon's Brigade of Guards commander, Colonel Joseph Garba with their plans and an Ironsi type ultimatum; He (Garba) could make it easy and bloodless, or difficult, that is bloody. But whichever way, they were going to succeed and he had a choice either to work with them or to try to stand against, and be consumed by events. The '75 coup, though eventually bloodless, has been described as the most intriguing in Nigeria's history, with a lot of twists and turns, including the eventual timing of the coup to when General Gowon was attending the OAU conference.

Just like in July 66 when he had to shuttle between the Ikeja Cantonment base of the coupists led by Murtala, and the Government's resistance led by Gowon, Garba was once again right in the middle between Gowon, the man he had sworn to protect, and Murtala, the man the coupist were trying to install.

There was a strange turn when General Gowon's plane suddenly returned from Uganda, but it was Gowon's aide-de-camp (ADC), Lt. Col. William Walbe, returning to get some documents for Gowon. Garba tracked Walbe to his home & with a truckload of soldiers outside, spoke to Walbe about their long association from their days as 'Boy Soldiers' in the Nigerian Military School, with a plea not to get killed by fighting the Coup[20].

---

[20] (Omoigui, Military Rebellion of July 29, n.d.)

Colonel Joe Garba, Col William Walbe

The coup eventually succeeded and General Murtala Muhammed, who was conveniently out of the country as the final phase of the coup played out, was allowed to fly back home (despite a closed Airport) to take the reins of power[21].

A final twist played out after the success of the coup as the plotters turned on their mentor, Murtala Muhammed, and demanded a power sharing arrangement, which in Nigeria usually means some form of direct access to the treasury. Murtala, who held the 'juicy' Communications portfolio under Gowon, was so upset about the demands for power-sharing and offices to be held[22] that he was willing to give up the offer to become the head of state. The intensity of that negotiation meeting seemed to have had a deep impact on Murtala Muhammed, as questions over identity and origins seemed to stir him to see himself as a Nigerian, rather than a swordsman of the foreign hegemony that has always sought control of the Nigerian State.

The reality of those negotiations however led to a complete clearance of the top echelons of the civil service, and expectedly, the military. This 'treasury accessing' clearance has been

---

21 (Omoigui, Murtala Ramat Muhammed (1938-1976), n.d.)

22 (Siollun, n.d.)

blamed [23]for the normalization of corruption and the disruption of the development[24] of the country. However this type of diagnosis, though correct at some superficial level, only addresses the symptoms and not the root causes of the deep issues facing the Nigerian State, as we have seen.

Murtala tried to be a unifying factor during his short time as head of state, forging an impressive non-aligned, pan-African identity and outlook for Nigeria, a radical change from the divisive and sectional stance he had taken in public life before that moment. He was particularly close to Colonel Ibrahim Taiwo (known as Ibrahim Kagara earlier in his military career) as both were devoted swordsmen of the hegemony who faced identity questions. Both were also accused of involvement in the Asaba massacre. Sadly, the fate of both men was tied in another tragic twist as they were felled by Assasin's bullets on February 13, 1976 during a coup led (or announced) by Lt. Col. Buka Suka Dimka.

23 https://www.vanguardngr.com/2015/08/how-corruption-hit-civil-service-asiodu/

24 https://www.sunnewsonline.com/murtalaobasanjo-stopped-nigerias-march-to-greatness-phillip-asiodu/

Gen. Murtala Muhammed and Col. Ibrahim Taiwo

1976 - In February of 1976, General Murtala Muhammed, Colonel Ibrahim Taiwo, and a few other officers lost their lives[25] in an attempted coup that was primarily carried out by middle-belt officers of Benue-Plateau origin against the regime of Murtala Muhammed. The coup eventually failed and the coup announcer, Colonel Buka Sukar Dimka was arrested in Abakaliki[26] and tried in Lagos. About 30 other officers and men of the military and the police and also a few civilians were executed for their roles in the coup[27]. General Illiya Bisalla, the Commissioner of Defence at the time, was also mentioned and implicated by Colonel Dimka in the coup, although his involvement, and that of some of the other executed officers, including Joseph Gomwalk have been questioned.

25 (Omoigui, Col. Dimka's Failed Coup Attempt, n.d.)
26 (Omoigui, Col. Dimka's Failed Coup Attempt, n.d.)
27 (The New York Times, 1976)

In most other circumstances, General T. Y. Danjuma would have assumed the role of head of state (Yar'Adua could have also made an aggressive and audacious move). However, as the coup was mostly carried out by middle-belt officers (to reinstall Gowon according to some, including the Government, even if this was unlikely), it was difficult to make T. Y. Danjuma, also from the middle-belt, the primary beneficiary of the coup that took out a strong representative of the Fuduyan Fulbe hegemony.

As such, the lot fell on General Olusegun Obasanjo, who was extremely reluctant and some say, fearful to take the position. Obasanjo bent backwards to try to placate the Hegemony, and one of his very first decisions taken was to award a double promotion to Shehu Musa Yar'Adua from Lt. Col to Brigadier General[28]. This was to make Yar'Adua the Chief of Staff Supreme Headquarters after rapidly bypassing dozens, if not scores, of his superiors in the army. Obasanjo continued to placate the Hegemony, primarily through the Bello-Ribadu officers, for instance, Muhammed Buhari, was made the Petroleum Minister (Federal Commissioner at the time).

Yar'Adua and Buhari had leading roles in the Murtala Coup and the Obasanjo Government

[28] (Omoigui, Military Rebellion of July 29, n.d.)

Shehu Musa Yar'Adua, with an aristocratic background as the son of the former Lagos Affairs minister in the first republic was undeniably ambitious, and it was obvious that Obasanjo greatly deferred to him and the other Bello-Ribadu Officers. As these things play out, there is never enough space at the top, and other Bello-Ribadu boys were on the outside, looking in (possibly with scorn). Sani Abacha and Tunde Idiagbon for instance, reportedly tried to instigate their GOC in Ibadan, James Oluleye, to oppose the coup once Shehu Musa Yar'Adua's leading role became obvious. Yar'Adua clearly attracted a lot of envy from his his colleagues, contemporaries and officers from the far north, and this was to catch up with him much later on.

For instance, Ibrahim Babangida, as military head of state inexplicably excluded and removed Yar'Adua from the political process of the late 80s and early 90s.[29], while Sani Abacha imprisoned him. To the rest of the country, most of these gentlemen were beneficiaries of the Ribadu-Bello illegality, so a couple of them had little reason to get too worked up by Yar'Adua enjoying extra privileges (like a double promotion) from the same illegality and nepotism, due to his aristocratic background.

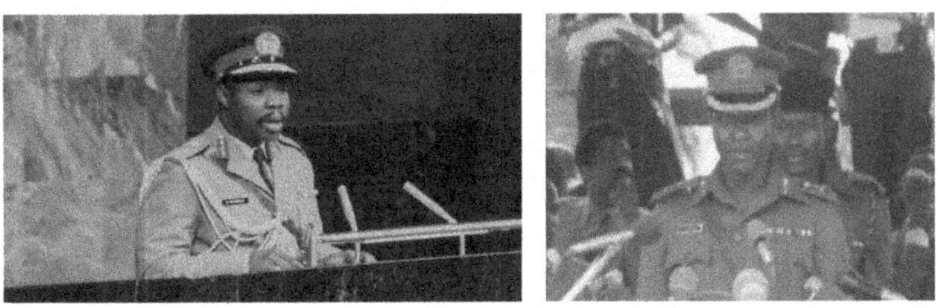

General Olusegun Obasanjo and Bri. Gen Shehu Yar'Adua

[29] (Oduyela S., 2004)

1979 - Obasanjo, in 1979, fulfilled his promise and handed over to a democratically-elected civilian government. T. Y. Danjuma has said that as they (he, Yar'Adua, and Obasanjo) were leaving the army, he made sure that the army was left in the care of Alani Akinrinade, who became Chief of Army Staff[30]. Akinrinade has been credited by many, alongside other sub-commanders (including Alabi-Isama and Tomoye) as playing crucial roles in ending the civil war by proactively reaching out to officers on the Biafran side when it was became obvious that the fighting dragging on beyond what was necessary.

The military left governance in 1979 and General Akinrinade was named Chief of Army Staff, to help safeguard the civilian government. Danjuma also wanted Akinrinade to be Chief of Army Staff because he knew that the latter could handle the indiscipline and aggressive tendencies of Bello-Ribadu Officers.

Obasanjo and Danjuma were glad to leave Office in 1979. This however made Yar'Adua's path to his ultimate objective a bit more complicated.

1980 - Fulbe Civilian President Shehu Shagari soon shot himself in the foot as he relieved Akinrinade of the command of the Army and 'demoted' him upwards, creating a new role called the Chief of Defence Staff. Akinrinade resigned from the Army shortly

30 (Daily Trust, 2019)

afterwards and was replaced by Gibson Jalo, whose humiliation was probably worse than Akinrinade's, as Shagari created the role of a Deputy Chief of Army Staff for Inuwa Wushishi. Jalo was relieved of his role after just over a year, and as such the men left behind (Southerner and wrong type of 'Northerner' in the books of the ruling clans) to contain the aggressive and unruly tendencies of the Bello-Ribadu crew had been pushed out by the man they were meant to protect.

Shagari, either listening to bad advice or taking bad decisions on his own, abolished the position of the Deputy Chief of Army Staff as soon as he got the 'right man' as Army Chief. He would soon get burned by the men his NPC bosses had forced into the Army in the 60's.

Shagari, Akinrinade, Jalo

1983 - Perhaps testing his limits of his 'command' structure, Buhari, then a Major General and GOC of the 3rd Armoured Division in Jos, defied the presidential order and invaded Chad, following incursions by Chadian troops into disputed border areas and islands around Lake Chad. The invasion almost reached the N'Djamena, the capital of the country. Shagari chose not to punish or reprimand his fellow Fulbe for the defiance.[31] An un-

31 (Omoigui, History of Civil-Military Relations in Nigeria (5): the Second Transition (1979-83, Part 2), 2021)

derlying reason for this may have been due to the fact that Hisen Habre's Toubou ethnic group have had a history of clashes with the Fulbe [32]across countries bordering the Sahara (Mali, Niger, Chad), and these key Nigerian State actors, were using Nigerian soldiers, weapons and resources to settle historical and foreign scores that mean absolutely nothing to indigenous Nigerians.

L: Brig General Ibrahim Bako with Major General Muhammadu Buhari, Head of State.

1983 - On December 31, 1983, Brigadier-General Sani Abacha announced that there had been a coup removing the civilian government from office. Abacha, a Kanuri raised in Kano, finally emerged from the shadows. Abacha's background made him close to Muhammadu Buhari, leader of the coup, and a key swordsman of the hegemony over the years, but without the aristocratic background of Yar'Adua. Buhari, though Fulbe, had a Kanuri mother, and like Murtala and Taiwo before him, seemed eager to prove his devotion to the cause.

Buhari's coup has been said to be bloodless and largely so. However, there were some intriguing deaths during the coup especially that of Brig. Gen. Bako, who was rumoured to have

[32] https://climate-diplomacy.org/case-studies/conflict-between-fulani-and-toubou-niger

had a close relationship with the new head of state at the time. Buhari also enjoyed a much laundered public image, thanks to the media influence of his older nephew, Mamman Daura.

1984 - 1985 - Despite its tough stance on indiscipline in public life, the Buhari administration proved to be clueless in changing the fortunes of the country and quickly plunged the country into serious economic crises. There were soon shortages of essential commodities, which then came under various forms of controls.

1984 - In another tough public stance against corruption, Buhari orders a change in the colors of all the Naira (fiat paper) currency ranging from ₦1 – ₦20 (between 1 naira and 20 naira) with a limited time for the public to change their currencies in banks. Among the reasons given was to pull the rug from under the feet of people that had stashed money abroad. Guardian Nigeria journalist Okunaba Adinoyi Ojo reported that key Fulbe Clanleader from Gwandu, and father of Buhari's ADC, Major Mustapha Jokolo imported 53 suitcases, and soldiers had prevented the customs from carrying out the Presidential directive of searching all imported luggage. (The Gwandu based clanleader is ranked 2,3 or 4 by Hausaland's Fulbe ruers, depending on who is doing the ranking with Sokoto, Kano & Kanuri Shehu of Borno, who has not identified with the Fuduyan Fulbes, pulled into the rankings). Future Vice President and Presidential Candidate, Atiku Abubakar, was the Customs Station Manager for the Airport at the time.[33]

Buhari would instantly have his pound of flesh on the Guardian, as two of its journalists were jailed for "false statements likely to bring the government or officials into ridicule or disrepute."

Mrs. Ransome Kuti and Prime Minister Tafawa Balewa

1984 - In a clear contrast to the Gwandu suitcase affair, Fela Kuti is arrested with foreign currency while heading out of the country for a concert in the United States and tried in what

[33] https://www.legit.ng/1102783-atiku-account-53-suitcases-saga-imported-nigeria.html

the Guardian described as 'sham currency smuggling charges' and jailed for 5 years.[34] The judge later apologized and said he had been ordered to jail Kuti[35]. Fela's fearless and witty renditions on the Nigerian State had been a source of irritation for successive Military dictators. A few years earlier, his sprawling compound 'Kalakuta Republic' was invaded by soldiers and destroyed. Everyone present, including his mother, famous activist Funmilayo Ransome-Kuti, was beaten and she later succumbed to her injuries. The order to invade the compound and get Fela 'dead or alive' was said to be right from the highest echelons of Nigeria's military dictatorship[36]. The Government commission of inquiry declared that it was the action of unknown soldiers.

L: Fela Kuti R: Customs Officer (later Vice President) Atiku Abubakar

1985 - Buhari's reaction to his inability to tackle growing economic hardship was to push the narrative of a disciplined

[34] https://www.theguardian.com/music/2004/aug/15/popandrock5
[35] https://www.spin.com/featured/fela-kuti-july-1986-interview-fela-freed/
[36] https://onlinenigeria.com/nm/templates/?a=12357

and uncompromising enforcer, with obvious examples being the jailing of Guardian journalists (Nduka Irabor and Tunde Thompson) and the retroactive implementation of laws to execute drug traffickers. Buhari has often failed to comprehend the illegality of the executions with responses suggesting that questioning the execution was akin to supporting drug trafficking, but clearly failing to see the danger posed by a Government convicting any one based on a law that did not exist (or did not carry the implemented punishment) at the time of the act/misdemeanor or crime.[37]

1985 - Given the human rights records and economic struggles of the regime, including shortages of essential commodities, the removal of the Buhari junta in August 1985 was generally welcomed. Babangida's bloodless coup was filled with intrigue and against the odds in many regards. A non-Fulbe, his ethnic origin has been the subject of much debate and rumors with Kano, Bida, Minna and Ogbomosho offered as answers to questions about his ancestral home. This was important for officers joining the Army in the early 60's, given the 'ethnic camouflaging' of a few of the young men to pose as mostly Hausa (much tougher to disguise as Fulbe or Kanuri, the other ethnic beneficiaries) to take advantage of the Bello-Ribadu policy.

It was significant that Buhari's Service Chiefs (Bali, Aikhomu, Alfa, with IBB moving up from being Army Chief and replaced by Abacha) were retained by IBB, suggesting that they all sanctioned the removal of their Principal, or worse, a complete disenchantment and lack of trust in his ability to lead.

---

[37] (Oredein, 1985)

1985 - IBB's rise to power was probably due to his personal charisma and ability to rally and relate with subordinates in the military. His role in quelling the Dimka coup in 1976 also gave him leverage with the Bello-Ribadu stronghold within the Military, which was strongly linked to the Fuduyan Fulbe clans.

At the core of IBB's rise to power were a number of Lt. Col's and Cols. at the time (known as IBB Boys) including David Mark, Lawrence Onoja, Tunde Ogbeha, John Shagaya, Anthony Ukpo and Lawan Gwadabe. Most of these Officers were from the 'Middle Belt' but IBB's reach extended beyond the zone, and the role of Major Sambo Dasuki, a key member of Hausaland's ruling Fulbe's in the coup is well documented. Dasuki served as IBB's ADC till he quit the position following a clash with Maryam Babangida, and was sent out of the country on a course to douse the tension. Dasuki also later served as Goodluck Jonathan's National Security Adviser.

Many of the IBB boys smoothly transitioned to civilian life after being removed from the Army by (civilian President) Obasanjo with David Mark, John Shagaya and Tunde Ogbeha serving in the Nigerian Senate.

Part of the public's charm with IBB was also due to his wife. Nigerian leaders had mostly been distant and unengaging, but suddenly, there was Maryam Babaginda who did not stay in the shadows, but was very much beside her husband in the public. Her 'Better Life for Rural Women' program (with no clear budgetary funding source) provided an independent platform for public engagement, apart from just being the wife of the Military ruler. Her Asaba origins were also a factor in the thought that IBB could be a leader for all people, given the tragedy suffered by the town during the civil war.

Mrs. Babangida and a foreign dignitary, Princess Diana

However, as has been seen throughout the perilous History of the Nigerian Military and State, rallying younger officers and making promises to them while seizing power has little to do with the good or effective administering of the country. Nigerians were eagerly looking for some sort of competent leader to relieve the seemingly endless rot and grind of the country over the past decades, but IBB's charm and ability to draw everyone close and make promises to just about everyone quickly turned into weakness and disaster.

The term 'North' in Nigeria is often synonymous with the interests of Hausaland's foreign Fulbe rulers, and the clan leaders soon set to remind IBB that he was a beneficiary of the 'Northern' magnanimity through the Bello-Ribadu policy. IBB quickly calmed this key power base by rejigging his cabinet [38]and ruling council and having Nigeria take up a full membership of the OIC in 1986.

---

[38] https://rainbownigeria.com/2020/11/28/ibrahim-babangida-regime/

1986 - Mamman Vatsa, friend and coursemate of Ibrahim Babangida (the latter was his best man at his wedding), notable poet, and the Minister of the Federal Capital Territory at the time was arrested with several other officers and executed for reportedly planning a coup. It was unique in that some Air Force officers were implicated in the plot and executed too. Many felt this was the path that started a long period of neglect of the Air Force[39]. And for over a decade, the graduating Air Force classes of the Nigerian Defence Academy did not go to flight school. This caused a generation gap of about a decade of pilots, which did not end until the restoration of democracy and Olusegun Obasanjo when the regular training of pilots restarted.

Dele Giwa, leading journalist and one of four founders of Nigeria's leading news magazine, Newswatch, was killed via a letter bomb in his Lagos residence.

L: Nigerian literary icons, JP Clark, Achebe and Soyinka visit IBB to plead for Vatsa's life. R: Vatsa and IBB in much happier times.

1987 - Ibrahim Babangida created two states shortly after taking office. Akwa Ibom and Katsina. He also created some more states in 1991 making him the only head of state to have created states twice within his administration. All states that have been created in Nigeria have been created by northern military

[39] (Adegbamigbe, 2020)

generals, except the creation of the Midwest state, which was aided by the deal between Akintola and the NCNC members of the Western House of Assembly during the Action Group crisis (1962-1963)[40].

1990 - A coup, announced by Tiv Major, Gideon Orkar attempted not just to overthrow the government, but also to excise the far north (Hausa and Kanuri countries), from the rest of Nigeria. It was the third time since 1966 that military officers had tried to carve out and forcefully partition the country. The first was the attempt by Col. Murtala Mohammed to remove the 'North', as he determined it, from Nigeria, and the second, was the Biafra war when Colonel Ojukwu tried to remove the south-east from the country. The Orkar coup was crushed, reportedly with the Chief of Army Staff Sani Abacha playing the lead role in quelling it[41]. The varied reactions to the Orkar Coup perhaps highlighted the sharp differences and divisions that exist in the country.

Perhaps the reaction or the writings of respected journalist, Kabiru Gwangwazo showed the schism that exists in the country on this matter. In a series of articles he has written on the coup, titled "Gideon Orkar's Madness," he mentioned that working at a media house at the time, all the journalists from the other parts of the country rejoiced at the thought that the coup was successful.

The feeling on the streets at the thought of the excision of the far north from the country certainly matched that in the Triumph newsroom with the general sentiment that the 'troublers' of the rest of the country had been excised. In his

---

40 (Diamond, 1988)

41 (Akinbode, 2020)

own words, *"For me, then a reporter/sub-editor, at The Triumph newsroom, a mini-Nigeria of a news hub, the unenlightened response to the Orkar lunacy by almost all of my non-Muslim colleagues, especially from the Middle-Belt and the Igbo East, in particular, was a painful eye-opener. I understand, for IBB too, it was an eye-opener."*[42] Gwagwazo seemed particularly pained that 'Middle Belters' were leading and celebrating the attempt to split Nigeria as he adds again shows the very unenlightened (using his own words), but extremely worrying perspective of many of the Hausa and Kanuri elites and intellectuals who seem determined to close their eyes to the issues staring at them right in the face, which is that Hausaland under foreign Fulbe control, and to some extent Kanuriland, when willing to enter into a partnership with the foreign Fulbes, has been a source of nothing but trouble, burdens and threats to the rest of the indigenous Nigerian countries have been forced into the Nigerian State with them.

Gwagwazo then continues and says the *'holistic and unified Northern Nigeria' finally and effectively came to an end. That was when that Northern unity was killed. IBB was obviously the main reason for that murder so gruesome. This is despite his being a principal beneficiary of the' 'Monolithic North bequeathed us by the North's Sardauna Ahmadu Bello, managed effectively and efficiently by our successive leaders.*

Without doubt, the Orkar coup attempt was madness and doomed to fail because the odds were stacked against it. Firstly, no body has the right to murder or execute anyone while unilaterally playing judge, jury and executioner whether it'd be Nzegwu, Dimka, Orkar or Murtala and the bulk of the Bello-Ribadu crew

---

[42] https://www.vanguardngr.com/2018/05/gideon-orkars-madness-another-april-anniversary-2/

who murdered over 200 of their colleagues in Ikeja, Abeokuta and Ibadan.

A coup done the 'right way' is a thankless, no reward, full risk endeavor as the plotters will have to arrest their targets without any bloodshed, and then hand them over to the police or command structure, while surrendering to the same authorities for insurrection. In effect, there is no right way to carry out a coup. That is why conversations about governance and societal arrangements must be thorough, frank and not suppressed. Nigeria's current constitutional arrangement which baselessly and hilariously comes to the conclusion that Nigeria is one indivisible nation (Section 2) with little consideration for its extremely shaky history and foundations, and with sections 8 and 9 making it virtually impossible to achieve the much needed constitutional change through the national assembly is however a catalyst for continuous agitation and demand for a dissolution of the Nigerian state to its forcibly conjoined constituent countries.

It must however be pointed out that no violent coup has succeeded (on its own) in Nigeria's troubled history, and once you throw violence and murder into the mix, the dynamics can quickly get out of control.

- In January 1966, Nwobosi arrested Fani-Kayode, but executed Akintola even after he surrendered, due to the initial resistance put up. Almost all the other plotters resorted to violence as the first resort. The takeover attempt was stalemated (depending on whether you think Ironsi was pulling the strings on Ifeajuna all along and stepped in as planned), and the coupists had to surrender.
- The unpunished acts of July 1966 were not a takeover attempt, but acts of war against what would soon be

a foreign country. These plans also ended in a stalemate, with Senior civil servants, Yakubu Gowon and the English Ambassador stepping in to end the imbroglio, and 'negotiations' rooted in selfish interests producing Yakubu Gowon to try to save the Nigerian state for reasons that have nothing to do with the good and benefit of indigenous Nigerians.

- The Dimka coup of 1976 was unsuccessful, with IBB earning 'brownie points' with the power base for his role in quelling the coup.

Coups in Nigeria go beyond a group of young men shooting their way to the TV and Radio house and killing a couple of politicians along the way. Almost nobody will hear a speech on NTA these days anyway and so maybe the plotters will upload a youtube video?

That is why the bloodless coups (especially Yar'Adua's on Gowon) have had the most intrigue, and even with IBB's overthrow of Buhari (which was more of an indictment of Buhari by all his top Generals, apart from Idiagbon), Babangida was quickly reeled back in to do the bidding of the power base that believes it owns Nigeria. The support of the top military brass was not sufficient for effective control of the country.

While the plan for Orkar's coup was madness and not grounded in reality, the sentiment behind the coup was very much sane, logical and in some ways predictive, as the states excised by Orkar, somewhat mirrored the states that 'removed' themselves from the secular Nigerian Nation with the declaration of Sharia law just over a decade later. It is certainly much saner than the logic that assumes that a nation set up by, for (the benefit of) and controlled by foreigners with no cultural or historical links to the indigenous peoples and ethnicities must exist as 'one indivisible nation' that takes the right and opportunity for self determination away from the indigenous peoples.

Orkar's coup speech (like all others) had some laughable and ridiculous aspects, and the forced removal of legal residents from any part of the Nigerian state must never happen (invading foreign AK47 wielding terrorists occupying forests reserves do not meet the criteria of legal residents). However, a demand in the speech was quite telling.

> *"To send a delegation led by the real and recognized Sultan Alhaji Maccido to the federal government to vouch that the feudalistic and aristocratic quest for domination and operation will be a thing of the past and will never be practiced in any part of the Nigeria state".*

Orkar was in essence requesting that the Fulbe rulers of Hausaland should be content with their 'prize' and stop using it as a launchpad to harass and bother the rest of the country. Seems sane and logical enough!

Getting back to Gwangazo, what is even crazier than Orkar's coup is to think that there has ever been a 'Monolithic North' and then have the confidence, not just to say it, but put it in print.

In addition to the administrative convenience of ensuring he could spend half the year social climbing in England, Lugard decided to add the 'middle belt' to the far North to achieve fiscal balance to a Hausa country that had been turned from one of Africa's richest to one of its poorest within 100 years of rule by the foreign Fulbe clans (this is common with virtually all forms of colonialism). I wonder how he could suggest the North was monolithic,

- When Ahmadu Bello reasoned that it was better (for him) to be dead than be Hausa. How monolithic was that north?
- Or when he mentioned that the Tivs were his grandfather's slaves? The appointment of an administrator (Aliyu Muhammed) and the use of the Army to subjugate the Tivs and eliminate efforts at self determination were among the reasons stated by the January '66 coupists for their actions.
- Prime Minister Tafawa Balewa certainly didn't think the North was Monolithic. Not when his grandmother told him every night as a child that he had a responsibility to 'kill the Fulani and drive them from our land' when he grew up.
- Maybe he is referring to the North, where even in the Fulbe ruled Hausland, the Maguzawa have been largely alienated within their own country and even somewhat stripped of their ethnicity, one of the most basic forms of identity.

If we limit the definition of monolithic to large, slow to change (it has actually rapidly retrogressed and is dragging the other countries in the Nigerian state with it) and add 'subjugated by foreigners' to the mix, then Gwangazo and others who love to repeat this 'monolithic north' line, may be right.

However, despite Lugard's gift to Hausaland's Fulbe rulers aligning with his need to social climb in England), and subsequent fraud and gerrymandering by the colonials for reasons previously stated, there has never been one Monolithic North. There is one Hausa Country though, and it is currently split across borders that were artificially created in Berlin. Little wonder why the proponents of the Monolithic North always seem to miss the elephant in the room while chasing phantoms.

1990 - Ibrahim Babangida moved the Federal Capital to Abuja, in the middle of Nigeria, mostly created from Gbagyi and Nupe countries. The process was started by Murtala Muhammed based on the results of the Akinola Aguda panel in 1975.

1990 - Ibrahim Babangida moved the transition to civil rule to 1993, from 1990 and launched two political parties, the progressive-leaning Social Democratic Party and the conservative-leaning National Republican Convention. He also banned a few people, including Gen. Shehu Musa Yar'Adua (Rtd.) from the political process[43].

1992 - IBB's regime also witnessed the sad event of the C-130 crash in Ejigbo shortly after takeoff from the Lagos Airport. Over 150 Nigerian military officers and 8 other officers from four African Countries (Ghana, Tanzania, Zimbabwe, Uganda) were killed in the crash

1993 - Businessman Moshood Abiola wins the June 12, 1993 elections conducted by Ibrahim Babangida but during the announcement of the results, the process was suspended, and the elections later annulled. In IBB's own words, he annulled the elections because *'They do not trust me. Without Sani, I will not be alive today; without the North, I would not have become an officer in the Nigerian Army and now the President of Nigeria I don't want to appear ungrateful to Sani; he may not be bright upstairs but he knows how to overthrow governments and overpower coup plotters.*

---

[43] (Oduyela S., Owners of Nigeria Part 4, 2004)

He saw to my coming to office in 1985 and to my protection in the many coups I faced in the past, especially the Orkar coup of 1990 where he saved me and my family including my infant daughter. Sani risked his life to get me into office in 1983 and 1985; if he says he does not want Chief Abiola, I will not force Chief Abiola on him'. [44]

By 'the North', IBB is of course referring to the foreign Fulbe rulers of Hausaland who adopt the 'Northern' title when it is time to exercise control over the Nigerian state, with religion sometimes used as the trump card. His stated indebtedness regarding the Army dates back to the Bello-Ribadu policy of the early 60's, the reason Nigeria's 'deep state' was able to reel him (IBB) in from about 1986.

For Abiola, it was the 2nd time his Presidential ambition had been truncated. In the 1979 - 1983 era, as a major backer and financier of the ruling party, he seemed to be under the impression that he would be the (NPN) Party's Presidential candidates at some point in the 80's, but since parties (and candidates) in Nigeria do not need outside private funding once elected, he was eased out of the party with those famous lines 'The Presidency is not for sale' by Shagari's Transport Minister, Umaru Dikko Abiola may have assumed that the banning of other heavyweights like Yar'Adua from the 1993 contest was a clearing of the path for him, but the rug was pulled from under his feet right at the finish line.

Abiola, an accountant, was the Nigerian head of ITT (a global Telecoms giant that was split up in the 90's and whose successor companies include The Hartford, Starwood Hotels and ITT Tech) who had struck up a close friendship with

44 https://www.thenewsnigeria.com.ng/2021/06/11/june-12-annulment-why-i-made-sure-abacha-also-ruled-nigeria-babangida/

Communications Minister (Commissioner) and later Head of State, Murtala Mohammed. Murtala was Abiola's access not only to Communications, but also to key Military officers, especially the Bello-Ribadu power base. Abiola also started his own telecommunications company (RCN), even though he remained on the global board of ITT. He also delved into other business ventures, including Media with the Concord Newspapers.

Abiola was criticized for his wheeler-dealer tendencies and personal excesses, with supporters responding that his personal life should be separated from public office and political ambition. While the personal affairs of public officers should be protected from the public space, there should be some accountability for personal conduct and demand for personal integrity and ethics. This is probably more necessary for Nigeria and its constituent countries or successor states, given the relatively weak systems and institutions. In addition to competence, those in charge of public affairs and the treasury must be men and women of integrity.

Anthony Enahoro (with Abiola) made a rare public appearance (at the time) to condemn the 'annulment' of the 1993 electoral process.

As such, nobody seeking public office should have more than four biological children. Nobody seeking public office should have more than one spouse[45]. Public officials should be models and examples, given the challenge posed by a rapidly growing population. This limitation should apply to all public life including civil servants, legislators and contestants for and holders of political office. Situations where judges, legislators, civil servants and public executives have children by the dozens and wives and concubines that cannot be counted on a hand isn't just a recipe for corruption in governance, but an entrenchment of poverty for the general populace. Nigeria or its successor states, should they remain secular, must ensure that all public officials adhere to the single spouse, four children limit policy.

Abiola may also have sealed his fate when he vowed to demystify the National Oil Company, NNPC, which itself is an alternate, or off-balance sheet treasury for Nigeria's rulers and owners. He promised transparency in this key, but unnecessary and value destroying source of government revenue[46]. Almost 30 years later, NNPC's revenue is still as convoluted as it has always been as the organization continues to destroy value, with billions of dollars lost each year between the clear and straightforward revenue generated (petroleum profits from crude sales) and the remittance to the public treasury.

45 The exception will be for the first wife in a polygamous marraige

46 Channels TV Youtube Video - MKO Abiola speaks about his plans, vision for Nigeria

L: Abiola with Babangida R: Abiola with Abacha

1993 - Sani Abacha removes the Interim National Government from the scene via a bloodless coup. This ushers in a period of crazed and brutal dictatorship by another Bello-Ribadu beneficiary.

Abacha, the man who announced the initial Buhari coup had been a workhorse behind the scenes for various military coups. He probably felt that this was his time and his reward for serving various military juntas.

Former US military chief Colin Powell said Sani Abacha's CIA profile was the most troubling he had ever read[47]. But this should not be surprising, given the background of these officers who were forced into the Army by Ahmadu Bello and Ribadu in the 60's. The tendency is not to value what you did not earn, and these young men, who were likely not qualified to be officers in the army ensured that Nigeria (at least the indigenous Nigerians, since the 'owners' tend to have an 'opposite people' (© Fela) mentality) paid a bitter price for Bello-Ribadu corruption of the Officer Corps of its Army.[48]

47 https://www.chicagotribune.com/news/ct-xpm-1998-06-09-9806090183-story.html

48 (Oduyela S., Owners of Nigeria: Part 6, 2004)

1993 - Sani Abacha appoints Allison Madueke as Chief of Naval Staff, the first Igbo service Chief since the end of the Nigeria-Biafra civil war. He was removed from office less than a year later.

Abacha and Babangida

1995 - Sani Abacha executes poet, playwright and writer Ken Saro-Wiwa under circumstances that remain controversial to date. Abacha and Saro-Wiwa had known each other either during or shortly after the civil war and Abacha believed that he had assurances from his old friend Saro-Wiwa that there would be limited (or no more) activism in Ogoni land. Thus, he felt betrayed by Saro-Wiwa's continued activism. Saro-Wiwa and 8 other activists were arrested and prosecuted on charges of the murder of other leaders in Ogoni land. Subsequent and independent investigation suggests that the murders were not ordered or committed by Saro-Wiwa and it was a setup[49]. Many

[49] https://www.independent.co.uk/news/world/africa/ken-saro-wiwa-was-framed-secret-evidence-shows-2151577.html

military officers, probably acting upon orders, played a very ignoble role in the Ogoni affair. Saro-Wiwa and his co-defendants were convicted by a special tribunal led by Justice Ibrahim Auta and notably, were executed even before the 30-day period that should have been allowed for an appeal of the conviction. The outcry, both local and foreign) over the executions was swift and massive.

L: Kenule Saro-Wiwa R: Nelson Mandela

1996 - Following criticisms from South African president Nelson Mandela[50], Gen. Sani Abacha crazily stopped the national team, the Super Eagles from attending the Nations Cup in South Africa. That generation of the Super Eagles was considered as one of the finest up to that point and were the defending champions of the tournament and would likely have repeated in South Afric, as shown by getting the Olympic gold in the same year in Atlanta. Nigeria was subsequently banned from participating in the two following Nations Cup tournaments, prematurely ending the national team careers of some of the finest generation of Nigerian players.

50 (Obiejesi, 2018)

Abacha stopped the Super Eaglesfrom defending their African title in South Africa in 1996, and were banned from the next Nations Cup. The Eagles were favoured to win, and won Olympic Gold in the same year.

The Air Force continued to suffer from a shortage of young pilots and a reported lack of training for existing ones, supposedly due to an unwritten policy following the suspected participation of Air Force officers, who were to use fighter jets to execute the Vatsa coup. This may have come back to bite the country and its leaders in a very personal way, as pilot error was suspected in a number of military crashes, including one that claimed the life of Abacha's son, and other young Nigerians from notable families

1997 - Shehu Musa Yar'Adua dies in prison. Perhaps frustrated that opportunities for his shot at Nigeria's Presidency seemed perpetually delayed, Shehu Yar'Adua launched a public campaign against Abacha's plans to transition from military head of state to a civilian president. Sani Abacha locked Yar'Adua in jail and sadly, he was never to get out again. Abacha had been adopted by all the political parties as the Presidential Candidate, and previously unknown personalities were suddenly driving well funded and coordinated mass campaigns and rallies for

Abacha. These included Yomi Tokoya and GESAM (General Sani Abacha Movement for Peaceful and Successful Transition) and Daniel Kanu and YEAA (Youth Earnestly Ask for Abacha). A couple of known and respected public personalities, including Daniel Amokachi, Danladi Bako, John Fashau, Nwankwo Kanu and shockingly Maitama Sule) also got involved in the Abacha for President effort.

1998 - Sani Abacha died unexpectedly a few hours after meeting PLO chairman, Yasser Arafat in Abuja. Abacha's CSO, Major Al-Mustapha suggested that Abacha was clandestinely poisoned by one of Arafat's security officers. Even though reports are murky and somewhat scandalous, Abacha is reported to have died later that night after hanging out with close pal, then FCT minister, General Jeremiah Useni[51].

Sani Abacha greets Yasser Arafat

The moments following Abacha's death sadly reflect how thin and shaky the foundations of the Nigerian State are, and

51 (Oduyela S., Owners of Nigeria Part 8, 2004)

why even the military's role as a basis for unity is flawed and distorted.

Abacha's Chief Security Officer (CSO), Al-Mustapha supposedly consults Ibrahim Babangida and other key 'Northern' figures and then calls a couple of senior officers (excluding Jeremiah Useni, who was the most senior officer, and had diligently fought for the foreign Fulbe hegemony (aka 'the North') since 'Aure Paiko', to the detriment of his own people, according to some. The main figures consulted to decide who was going to become the next head of state were General Ibrahim Babangida and then Lagos administrator Buba Marwa (reportedly son of one last set of 'Glover's Hausas type soldiers (born in Lagos, but notably 'pro-hegemony'). Abacha's wife supposedly favoured Marwa, showing how a few people think they can (and do) decide the destiny of the Nigerian State because they have access to brute force and weapons to kill and intimidate indigenous peoples.

Al-Mustapha has subsequently stated 'they' had boys ready in Abuja and in surrounding places like Keffi to take over the country and could have, if he wanted, started a 'revolution' (He is obviously not a student of contemporary Nigerian History). He was reportedly talked out of doing so by Babangida.

Al-Mustapha also said that he considered shooting Abdusalami Abubakar, the man who was eventually selected by the group to become the head of state at a 9:00 am meeting the following morning, because Abubakar dared to sit on the chair of the late Sani Abacha.

The announcement of Abacha's death later that day set off scenes of instant and unplanned jubilation and street partying in the Middle Belt and the South of the Country, with Fulbe controlled Hausa, and Kanuri countries naturally taking the opposite stance.

Al-Mustapha, Jeremiah Useni and Abdulsalami Abubakar

1998 - About a month after Abacha's death, Moshood Abiola, who had been in detention for most of Abacha's regime was also reported dead in circumstances that have caused much speculation. Many felt it was an equalization and a settlement or an opportunity for a clean break from the past to give the country a fresh start. Abiola reportedly took ill during a meeting with senior American diplomat and one time Ambassador to Nigeria, Thomas Pickering. A heart attack was stated as the official cause of death.

1998 - 1999 - The government of Abdusalami Abubakar proved to be truly transitional and set an aggressive timetable for itself to leave governance. Abdusalami is often blamed for bequeathing the 1999 constitution to the country, which a lot of people called a fraudulent and faulty constitution, but the truth is that Nigeria is a conglomerate of several countries and it cannot and should not have one constitution if it is to survive. Perhaps the biggest knock on Abdusalami's government was the 1999 constitution that he bequeathed to the country.

Gen. Abdusalami just did what he could do to manage a really bad situation and left the scene as quickly as possible. He wasn't expected to fix any of the deep flaws or foundational issues that the country has always faced.

1999 - At the end of Abdusalam's transition process, former military ruler Olusegun Obasanjo eased past Architect and Shagari's

Vice President, Alex Ekwueme in the PDP primaries, and then defeated Economist and former finance minister Olu Falae in the main elections. The answer to the electoral question was known to most before the test, and it was believed that Obasanjo's emergence was to compensate the 'South-West of the country, for the presidency that was denied to Moshood Abiola.

L: Ekwueme and R: Falae lost to Obasanjo at various stages of the 1999 elections

Obasanjo and Abiola hailed from the same town, Abeokuta (even if there are interesting Anambra suggestions regarding Obasanjo) where they were students at Abeokuta Grammar School at the same time. While he could not be described as a 'Fuduyan Swordsman' like the Bello-Ribadu group in the Army, he had certainly posed no opposition to the cause throughout his time as an officer in the Army, including as Head of State, and was definitely a much safer choice than his major contenders. According to Ekwueme:

> *'Obasanjo was in Yola Prison, he had no idea how we formed the party and what it was formed*

> *for, but they foisted him on us as a President and leader of the party,*[52]

In Obasanjo, there was alignment between external powers and influencers, and Nigeria's internal foreign colonizers, who had gone too far, even by their own aggressive standards. It also appears that the Americans were directly interested in Obasanjo's candidacy, rather than the usual approach of leaving West African affairs to France and England. Obasanjo was in the good books of the Americans for his 1979 handover, and was close to Andrew Young, the former US ambassador to the UN, and maintained friendships with Jimmy Carter, the former US President and Walter Carrington, the former US ambassador to Nigeria.

With the amount of money available in Nigeria's shadow (Off Balance Sheet) treasury, it was not difficult for those who foisted him on Ekwueme and Co. to fund his campaign and steamroll him to the Presidency.

Obasanjo takes over from Abdulsalam

52 https://www.vanguardngr.com/2017/11/ekwuemes-words-marble/

Atiku Abubakar, Obasanjo's choice for Vice President was an interesting one. The post-Customs Atiku became the deputy leader of PDM (People's Democratic Movement) a platform (as opposed to a Party) that Yar'Adua had set up to actualise his long standing Presidential ambition. Atiku had just won the election to become Governor of the curiously named 'Adamawa' state, but gave that seat up to become Obasanjo's Vice President.

Atiku and Yar'Adua

Upon becoming president, one of Obasanjo's first moves was to retire all military officers who had held political positions; officers who had been ministers and military administrators or governors were retired from the military. This move was retroactively seen as probably saving the country, or at least keep it trudging along for another decade or so, until the fundamental issues return to the front burner. There was a lot of disaffection within the barracks at these(dollar) millionaire officers who combined wealth with ambition and military power. The decision to remove these political officers helped restore some camaraderie within the military and closed the chapter of the direct impact of the Bello-Ribadu stain on the Army.

1999 - Revelations during the Human Rights Commission (Oputa Panel), during accusations and counter-accusations by Abacha's military intelligence chief, Brig. Gen. Sabo and Chief of Army Staff, Ishaya Bamaiyi, and the chief security officer, Al-Mustapha revealed the depth of the brazen corruption, autocratic rule, and depravity of the Abacha regime. Abacha's government was notorious for crazy economic policies including issuing multiple official exchange rates. He was also notorious for parallel governance structures like the non-transparent and nepotistic petroleum trust fund, headed by Muhammadu Buhari who has insisted Abacha was not corrupt[53], despite over $4.6 billion [54] in stolen funds recovered. Expectedly, several key actors in clear corruption cases remain active in Nigerian public leadership as at 2021

L:Gen. Sabo, C:Maikyau (SAN) R:Gen. Bamaiyi Maikyau's cross examination of Sabo was one of the highpoints of the Oputa Panel, revealing an underbelly of the top echelons of Nigerian leadership that most criminal organisations would be ashamed of.

1999 - 2001 - Extrajudicial killings are reported by the military under Obasanjo in Odi and Zaki-Biam[55].

53 https://www.vanguardngr.com/2016/05/abacha-never-stole/
54 Daily Trust, Feb 16, 2020
55 (Watch, 2002)

2000 - 2001 - A legal partitioning of the country sees 12 states implement Sharia law. The States roughly matched the those that were announced as excised during the Orkar coup attempt and roughly matched the sections of Hausa and Kanuri countries within Nigeria. The ensuing riots, especially in Kaduna left thousands dead.

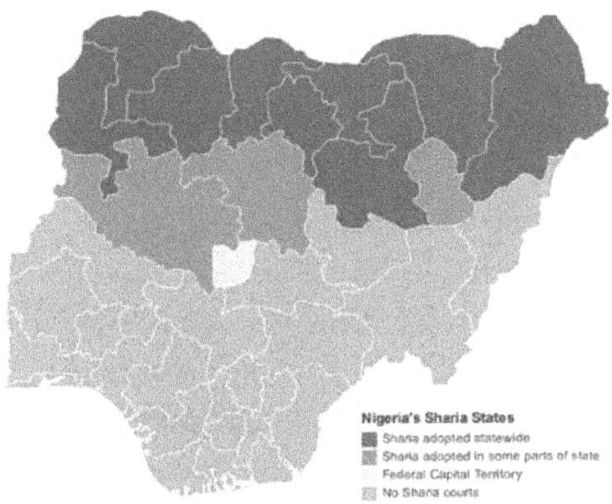

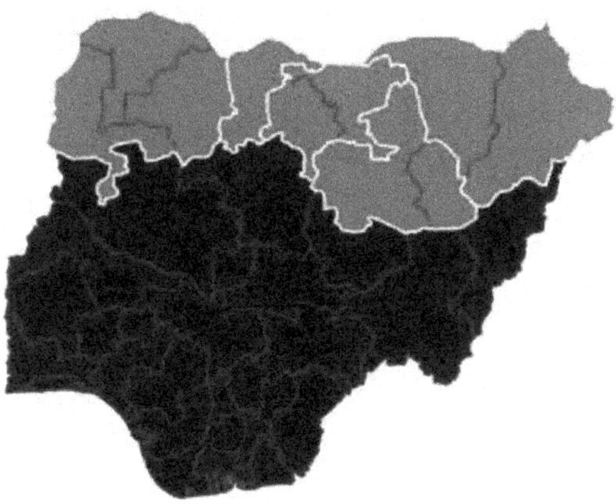

Orkar's announced partitioning/Excision in 1990

2001 - Obasanjo famously tried to sidestep the sharia issue and did nothing about it even though he was aggressive towards other constitutional violations such as the creation by Lagos State of LCDAs (Local Council Development Areas) alongside the Local Government Areas stated in the single Constitution forced on the various countries that constitute the Nigerian Nation. Obasanjo suspended the allocation and remittance of funds from the federation account to Lagos state.

2001 - The GSM revolution, which was said to have added at least 5% to Nigeria's GDP also happened during Obasanjo's regime. Little or no credit to the government on this though, as it just earned between $200 million and $300 per licenses from private Nigerian entrepreneurs, who collaborated with foreign partners to build the GSM network. Before the GSM telecoms licenses, there were less than 500,000 phone lines for over 120 million Nigerians at the time.

2005 - Obasanjo was also able to put a team together led by Finance Minister, Ngozi Okonjo-Iweala and head of the newly created Debt Management Office, Mansur Muhtar, to get a debt relief package and help Nigeria exit from the debt trap it was in due to the corruption and fiscal indiscipline of various military rulers in the past. Over $35 billion was owed mostly to Paris Club creditors by 2004. The relief included both write off and buyback of debt.

By 2021, Nigeria was back to a debt trap that was worse than the previous situation where the worry was that debt interest payments were more than capital expenditure. In 2021, debt interest payments have almost matched government revenue, with capital expenditure not even in the conversation. The significant difference (for good and bad reasons) is that the significant portion of the current debt is domestic, which mostly includes the pension assets of workers.

2006 - 2007 - Following disagreements with his Vice President, Atiku Abubakar, Obasanjo surprisingly selects Umaru Musa Yar'Adua, the independent-minded brother of the late Shehu Musa Yar'Adua as his successor, neutralizing and placating the somewhat formidable camp of the late Yar'Adua at the same time.

Obasanjo, a little older, presumably wiser and more experienced than in his previous stint as Head of State, still left office with the reputation of 'knowing what to do, but either side-stepping the issue, applying band-aid solutions or leaving the mess for another person to deal with. He failed to address the two of the most obvious forms of corruption in Nigeria, from which the perpetrators earn billions of dollars a year, both from the monetary (forex)and fiscal side (petroleum subsidies) of the economy. Both cases were classic examples of concentrated benefits to the powerful and corrupt, with the dispersed costs borne by the weak, powerless and often uninformed.

Nigeria's import dependent economy means that the demand for forex is almost always more than the supply which is from two main sources, Diaspora remittances and Crude Oil Sales.

The decades long clear and obvious corruption with remittances, netted hundreds of million of dollars to a couple of billions per year (simple arithmetic) for the perpetrators with the Central Bank legitimizing this fraud through the Outward Money Transfer Service. In addition to the ridiculous transfer fees, this costs Diaspora Nigerians and their relatives at home possibly up to a billion dollars most years.[56]

The fact that a 20% - 30% margin has consistently existed for decades between the street and official exchange rates, and that Nigerians in every major city know just where to go meet the 'Aboki' (Hausa word for friend, which is usually the first word uttered by the street money changers) to buy or sell foreign currency shows the scale and primary beneficiaries of the fraud.

There are many angles and aspects of the massive corruption in the Oil industry, but let us focus on a small subsection of the downstream sector with refineries. Nigeria's refineries lose over a billion dollars in most years[57], despite refining almost no products[58]. These losses represent real money from contracts[59], wages etc for no productivity, which is probably the purest form of corruption you can find This is obvious to anyone with a moderate intelligence, but the National Oil company also has the huge fringe benefit of providing a hidden and shadow treasury for the top echelons of Nigeria's ruling class, and those who believe they own the country. So, the status quo remains the same when just closing the refineries will save the country these losses (billions) a year, and better still selling them will even bring in some revenue, while saving the country from additional costs of foreign refining and the foreign exchange implications. The jobs and other multiplier effects would have been massive.

---

[56] https://www.vanguardngr.com/2019/10/26bn-diaspora-remittances-where-are-the-dollars-2/

[57] https://www.premiumtimesng.com/business/business-news/398151-three-refineries-lost-n1-6trillion-in-5-years-says-nnpc-audit-report.html

[58] https://www.thisdaylive.com/index.php/2019/12/17/nnpcs-refineries-record-losses-for-nine-consecutive-months/

[59] https://www.thisdaylive.com/index.php/2019/10/18/report-nnpc-spent-396-33m-on-refineries-tam-in-four-years/

In what critics describe as the classic Obasanjo style, he seemed to want to have it both ways, as he was clearly aware of the need for action, but waited till he was leaving office to announce the sale of the refineries[60]. Yar'Adua, perhaps not fully aware of the issues, reversed the sale[61] under pressure from Labour Unions that are sometimes teleguided by the National Oil Company.[62]

In October 2009, Yar'Adua eventually decided not just to sell the refineries, but stop the National Oil company from the entire downstream sector, in what would have been one of the biggest blows to corruption in Nigeria's history. The National Oil company begged for a month to put things in order before the announcement was made publicly. Yar'Adua never got the opportunity to make that announcement as he was flown out of the country reportedly in a comatose state, within that "extra month."

So Obasanjo left office with no dent to the biggest and most obvious forms and evidence of corruption in Nigerian society, but was happy to introduce useless ineffective and budget draining, role duplicating band aids like the EFCC and ICPC, effectively good for nothing more than hounding political opponents, and forcing the cross carpeting of strategic politicians in areas where whatever 'ruling party' of the day is relatively weak and intends to make inroads.

---

60 https://www.vanguardngr.com/2015/07/how-yaradua-was-pressurised-to-revert-refineries-sold-to-dangote-obasanjo/

61 https://punchng.com/yaradua-was-wrong-to-have-reversed-sale-of-refineries-by-obasanjo-ex-nnpc-executive-director-ihetu/

62 https://www.reuters.com/article/nigeria-refineries-idUKL194 4935220070719

Olusegun Obasanjo and his successor and PDP flag bearer, Umaru Musa Yar'Adua.

2007 - 2015 - Even though Umaru Yar'Adua had never been in the Military, there was still a sense that the Military still cast a long shadow on Nigeria's leadership and politics. Obasanjo's handover to the brother of the late Shehu Musa Yar'Adua, was clearly to spite Atiku but also to compensate for Shehu's death under Abacha. In 2015, former military ruler, Muhammadu Buhari was elected after defeating incumbent President Goodluck Jonathan.

Umaru Musa Yar'adua, Goodluck Jonathan, Muhammadu Buhari

2007 - 2010 - Umaru Musa Yar'Adua's health was a concern for most of his presidency and even before then but he tried to implement some aggressive reforms, famously removing the corrupt subsidy regime on diesel and kerosine and had ordered the removal of the petrol subsidies before he was taken out of Abuja under circumstances that remain questionable.

2009 - A major security development, and some would say resurfacing of structural and foundational issues during the Yar'Adua presidency was the emergence of Boko Haram, led by Muhammed Yusuf and with strong connections to the religious affairs commissioner in Borno state at the time. Since Nigeria was cobbled together (and before actually), there have been a prevalence of individuals and groups claiming legitimacy based on plans to govern society by a purer form of religious rules and edicts, so there was nothing that 'unique' about Boko Haram.

However, following the spate of clashes with the police and other security agencies, which resulted in hundreds of deaths, Boko Haram leader Muhammed Yusuf is arrested by the military and handed over alive to the police. Yusuf would be extrajudicially killed[63] and the group goes underground and starts an insurgency. The group and its like-minded religious terrorists soon start a campaign of bombings, initially against churches in the northeast but even in places like Minna, Niger state, but eventually spread take their campaign of terror on all and sundry[64]. Unlike previous times though, including the Maitatsine and various other religious uprisings in the north, the military has been unable to quell the Boko Haram uprising and up to 350,000 people have been killed and millions displaced as at 2020 according to the UNDP.

---

63 https://face2faceafrica.com/article/how-350000-people-have-been-killed-due-to-boko-haram-scourge-in-nigeria

64 (Walker, 2016)

Mohammed Yusuf and Abubakar Shekau

2010 - In another troubling moment for Nigeria's military, the capital, Abuja is taken over by troops as Yar'Adua is smuggled back into the country in unclear circumstances in which no one was sure whether he was alive or dead. Acting President Goodluck Jonathan was not conferred with, and the capital and villa were placed on lockdown as troops took over the city to return the president to the country. There were fears that a coup was in progress. The president was not seen in public until he was declared dead on May 5, 2010.[65]

NIGERIA'S GOODLUCK?

Goodluck Jonathan's Presidency may be viewed as a time when the Military (past or present) influence was not right at the helm of Nigeria's leadership, but it can still be examined from a Military perspective. Besides, ignoring this critical period will break the flow of our examination of Nigeria's history.

Yar'Adua was unable to officially transmit power to his Vice President, Goodluck Jonathan, as he was taken out of Abuja in unclear circumstances. As the days became weeks and months, with no President, and the Vice President unable to exercise

[65] (Ekine, The mysterious disappearance and death of Nigeria's President, 2010)

the powers of the President, the country was drifting into an unneeded crisis. This was where Bukola Saraki, Doctor turned politician and chairman of Nigerian Governors forum, stood tall in a uniquely Nigerian brand of statesmanship.

Saraki rallied all the Governors to put pressure on the senators from their States, and a few days later, the Nigerian Senate passed a resolution making Vice President Goodluck Jonathan, the Acting President, through the so-called doctrine of necessity, even without the official transmission of power from the President. There was a price tag, however, and a $4 billion withdrawal from the Excess Crude Account right after Jonathan assumed Presidential powers, only turned out to be an initial down-payment.

The Excess Crude Account (ECA) was a simple, yet effective mechanism to save the extra earnings from Oil proceeds over the Budgeted or Benchmark crude price for the year. However, there wasn't much in terms of safeguarding the monies saved, outside of a President's will and resistance (typical Obasanjo half measure, some would say). Smelling blood, the Governors demanded more[66], and just a few months later, the ECA hardly built up any substantial savings as withdrawals of the income became regular, a quiet transition from a savings to a current account[67]

This subtle but clear shift to the Governors as the fulcrum of political power was a key factor of Jonathan's eventual 5 year Presidency, but first there was the curious matter of James Ibori, who along with Saraki helped influence the emergence of Jonathan as Yar'Adua's Vice-President.

66 https://www.premiumtimesng.com/news/headlines/204809-pressured-squander-18bn-excess-crude-saving-jonathan.html

67 https://www.ft.com/content/0a7935ae-ca97-11df-a860-00144feab49a

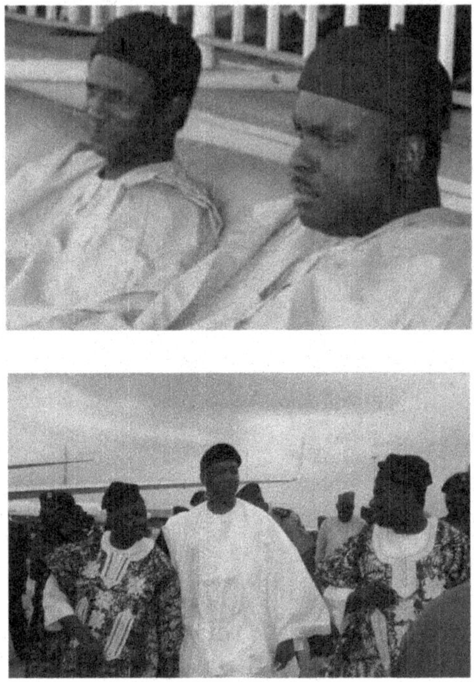

Ibori (with Umaru Yar'Adua) and Saraki (with Jonathan and Sambo) were foiled in their quest to be the new 'Capos' of Nigerian Politics.

While the Ibori-Saraki support for Jonathan had as much to do with neutralizing Peter Odili's path to the Vice Presidency (and to a lesser extent, Donald Duke's), the speed with which Jonathan turned on Ibori surprised most, and after a failed attempt to arrest him (Ibori) at Oghara[68], he slipped out of the country. He was eventually arrested in Dubai and extradited to England, where he was jailed on corruption charges.

68 https://www.reuters.com/article/ozatp-nigeria-ibori-20100422-idAFJOE63L02W20100422

Ibori, had somehow left himself exposed by not joining the Nigerian senate, which is the standard move of most ex governors in Nigeria. Perhaps he assumed his role in the emergence of Yar'Adua and Jonathan would guarantee his protection. It did, while Yar'Adua was alive and well. Saraki would not repeat Ibori's mistake and promptly ended up in the Nigerian senate upon completing his 2nd term as Governor of Kwara State.

Goodluck Jonathan undertook a number of other ultimately unwise political battles. Perhaps this was to curtail the growing influence of the Governors, or maybe it was to grow his own political base. These would later prove quite costly, but first of all there was the 'Banana Peel' of the 2011 election.

Hardliners like Lawal Kaita[69] threatened to make Nigeria ungovernable for Jonathan if he contested the 2011 elections, while other politicians like Adamu Ciroma asked him to respect the PDP's rotational Presidency agreement and not seek the party nomination.[70]

Widespread cases of underage voting have been reported in Nigeria's far North

69 https://pmnewsnigeria.com/2010/10/08/group-wants-kaita-tried-for-treason/

70 https://www.sunnewsonline.com/wanted-northern-president-2011-instead-jonathan-adamu-ciroma/

Despite these threats, Goodluck Jonathan defeated Buhari as expected in 2011. The reactions to the election exposed another dark and bitter aspect in Nigeria's history and politics. Spontaneous violence broke out in Nigeria's far North as the results were announced. At least 10 Youth Corps members from the 'South' and 'Middle Belt' were among the up to 100 people[71] murdered as senseless violence broke out over elections that Buhari definitely had no chance at winning.

It is normal to have Corps members assist the electoral commission with administrative and field functions during elections in Nigeria, and despite widespread criticism on the continuation of the Youth Service scheme in its current form, 'Corpers' are generally seen as 'other peoples' children that should be protected, wherever they are in the country. The killing of male and female corpers over elections where Buhari's supporters were proven to be involved in massive acts of rigging and underage voting, was a new low in the already depressing history of Nigeria.

The murdered Corpers included Chibuzor Ukeoma, Jehleel Adeniji, Tosin Teidi, Chukwuonyerem Nwazema, Seun Adewunmi, Obinna Okpokiri, Sule Akonyi, Elliot Adowei and Ayotunde Gbenjo.

Jonathan hosted the families and promised to bring the culprits to book. He donated N5 million (about $25,000 at the time) to each family, and promised automatic employment to someone from each bereaved family. [72]

The father of one of the slain Corpers, Mr. Awuchewu Okpokiri, captured the mood of most reasonable and peace loving people, when he observed that "communication in the

71 https://pmnewsnigeria.com/2011/04/21/northern-riots-100-bodies-recovered-jonathan-orders-probe/

72 https://www.vanguardngr.com/2011/05/post-election-violence-n5m-for-each-slain-corps-members-family/

last violence was nothing to write home about. A lot of us did not know (about) the death of these children until April 24 for something that happened on April 18. We also want to ask just like every other Nigerian is saying today that the NYSC scheme should be reviewed in all its ramifications."

There was no record of any commiseration with the families or regret over the killings by Candidate Buhari. Rather, he seemed to threaten more bloodshed with a troubling statement concerning the next elections about a year later that 'Dogs and Baboons will be soaked in blood'[73].

It is also not known if the Jonathan administration fulfilled its promises to the families of the slain corpers, outside of the monies paid, as 10 years later (2021), Sunday Dare, Buhari's Minister overseeing the NYSC scheme re-announced the employment of the family members as announced a decade earlier.[74]

[73] https://www.vanguardngr.com/2012/05/2015-ll-be-bloody-if-buhari/

[74] https://www.thisdaylive.com/index.php/2021/06/16/buhari-approves-jobs-for-2011-post-election-violence-victims-families/

Youth Corpers on electoral duty: At least 10 Corpers were hunted down and killed in the communities where they were posted to and served, as INEC announced the result of elections Candidate Buhari had no shot of winning.

The widespread violence that followed the 2011 election was insightful in a few ways. The first was that support for Buhari, even though very much in the minority across the Nation, was significant in Hausa Country. The CPC candidate, given his past record of incompetence, was seen as some sort of comic relief, each electoral campaign cycle. Statements credited to him, which suggested increasing support and defense of terrorism and extremism also meant that there was no way he could be accepted into the mainstream of Nigerian politics, even in its broken state on his sectional CPC platform.

However, the millions of out of school and itinerant (Almajiri) kids in Hausa and Kanuri Country would soon grow and have voter's cards and be a voting base. Hundreds of thousands, and possibly millions of them, already voted in past elections anyway.

Former Lagos Governor, **Bola Tinubu** pondered on these facts, and saw a window of opportunity. His limited, but well-funded Action Congress of Nigeria controlled Lagos, and occasionally, some surrounding states. It was a long shot, but still a window; and Jonathan, with his unwise political battles, would help expand that window.

Goodluck Jonathan was meant to be the one that was different. The one who came from nowhere and somehow made its right to the top due to a number of extraordinary events that catapulted him to the leadership of the Nigerian State.

Before Yar'Adua's death and Ibori/Saraki influencing Obasanjo to get him the Vice Presidency, he was Deputy Governor of the Ijaw heartland state of Bayelsa, but the Governor, retired Air Force Squadron Leader Diepreye Alamieyeseigha was impeached, leading to a relatively short tenure as Governor of Bayelsa before becoming Nigeria's Vice President.

Alamieyeseigha reportedly selected Jonathan as Deputy Governor for the same reason Ibori and Saraki wanted him as Vice President, someone who ticked some checkmarks, but would pose no threat or resistance to his or their own plans and schemes. His 'real-world' experience was reportedly largely limited to OMPADEC, the predecessor institution to the Niger-Delta Development Commission, before Politics found him.

There was a feel-good story to Goodluck Jonathan. A genuine outsider from the Niger-Delta region that had borne the environmental and sometimes oppressive burden of producing the oil that had been one of the weak glues binding the Nigerian State over the last few decades. He was not part of the political or economic elite and often told of how he grew up so poor that he could not afford a pair of shoes (hence the "Goody no shoes" nickname). There were hopes that his own background and outsider status would translate to massive improvements in the lot of the tens of millions of the extremely poor in the country.

However the lack of experience, indecisiveness, and a tendency to take on needless political battles were a huge constraint on his administration and would cost him severely.

His 'war' on Rivers Governor Chibuike Rotimi Amaechi was a clear example. The root cause of the issues is the subject of some speculation. It could have been:

1) To loosen the hold of the Governors that ran Jonathan ragged during his first months as Acting President, and later President. Amaechi took over from Saraki as Chairman of the Governor's forum.
2) Due to Jonathan taking sides in an internal State battle between Amaechi and Nyesom Wike, his former Chief of Staff, who was Jonathan's Minister of State for Education. Seizing control of Rivers' politics tends to be violent, even by Nigerian standards (hence the unpleasant nickname 'Rivers of Blood', and soon the State Governor was being hounded and harassed within[75] and outside his base [76]by 'Abuja power'[77]
3) Possibly due to a reported clash between Amaechi and Jonathan's wife and first lady, Patience Jonathan, a Rivers State indigene. Demolitions, Urban redevelopment, contracts and other stated reasons did not seem serious enough for the very messy, public and real

---

75 https://www.vanguardngr.com/2014/09/amaechi-laments-harassment-supporters-police/

76 https://www.modernghana.com/news/470735/amaechi-cp-mbu-and-politics-of-security-in-rivers.html

77 https://www.thecable.ng/amaechi-arrested-detained-in-ekiti

skirmish. Goodluck Jonathan himself suggested that this was the real problem.[78]

Jonathan's failure or oversight in reconciling the warring factions in Rivers (sadly, it always boils down to sharing political office and access to the treasury, as the departing Governor usually wants to install a stooge and go to the Senate) meant that Amaechi was feeling isolated within the ruling party, and became a prime target for the previously fragmented, but suddenly consolidating opposition.

Jonathan then stalled on implementing decisions that Yar'Adua had taken to end corruption in the downstream petroleum sector, perhaps with an eye on the 2011 Presidential elections. That was no excuse however, for his hilarious decision to put the fox in charge of the hen house and have the NNPC lead a supposed deregulation of the downstream sector that did not involve the removal of critical downstream sector assets including the pipelines and refineries from the control of NNPC and its subsidiaries, but seemed to zone in on Price increases and subsidy removal. The NNPC-Labour Union collaboration was in full swing once again with street protests, some of which were headlined by ignorant and misguided opposition figures. The President quickly ended the half-hearted process in January 2012.

A comprehensive reform in the downstream sector would have given Jonathan his signature economic achievement. Obasanjo could point to the jump in GDP from mobile phones (again the Government did almost nothing, but earn license fees here), clearing the external debt, and getting the military firmly

---

78 https://punchng.com/amaechi-had-problem-with-patience-not-me-jonathan/

out of the political space, even though he sidestepped a clear constitutional violation like Sharia. Implementing Yar'Adua's downstream sector reforms could have been a major boost to the economy in terms of jobs, income from sales and concession of downstream assets, forex savings (no more importation of refined product necessary) and possibly earnings, a boost in Government revenue due to the end of fraudulent deductions by NNPC, and if combined with forex reforms, would have definitely led to cheaper petroleum products regardless of the global crude prices or the exchange rate.

The reversal on the downstream sector reforms was not the only time that Jonathan changed his mind on critical and non-critical policy issues. He was famous for doing that. The University of Lagos (Unilag) became Moshood Abiola University (Maulag) and following protests, quickly transformed back to Unilag.

With the selection of Pre-Shipment Inspectors for crude oil exports, a Ministry of Finance panel had shortlisted four companies based on the pre-set criteria from the State House/ President's Office and concluded the process. A few months later, names of 11 companies were sent back with the suggestion that the process be redone. Eventually, all 11 were selected as Crude export Pre-shipment inspectors, with the contracts renewed as the President was about to leave office.[79]

A badly run orphanage with good, well intentioned people has a high probability of doing major damage and hurting the children, and a well and efficiently run criminal organization may be useful, under the right combination of close monitoring,

---

[79] https://www.vanguardngr.com/2017/05/jonathan-awarded-pre-shipment-inspection-contracts-72hrs-exiting-power-contractor/

risk management and threat of sanctions, as was the case with Lucky Luciano and the US Navy in World War II.

The scale of miscues from Jonathan's office during his Presidency suggests a poorly run State House, and the blame must naturally go to his Chief of Staff. There were suggestions

chose him as his most important aide, based on their similar experiences as 'long-suffering' deputies to Governors, with Oghiadomhe serving as a two-term deputy Governor to Lucky Igbinedion in Edo State.

The President's relative lack of experience, and possibly knowledge about the Nigerian State needed a strong office that would strengthen and guide, and not weaken the President. The key and relevant factors had to be considered before decisions were made. The number of decision reversals on even relatively simple issues suggest that the office did not provide the much needed support to the required level. Oghiadomhe was eventually replaced in 2014, but it was probably a little too late at the time.

Jonathan was still favored to win the election, but the opposition was no longer made up of fragmented units led by extremists and local champions who could be dismissed with a wave of the hand.

2010 - Goodluck Jonathan appoints Azubuike Ihejirika as Chief of Army Staff. Nigeria's first Igbo Army Chief since the end of the Nigeria-Biafra Civil War.

Madueke, First Igbo service Chief, and Ihejirika - first Igbo Army since the end of the Nigeria-Biafra Civil War.

2011 - Boko Haram attacks the UN building in Abuja, Nigeria's

group's strategy and capacity. This new dimension to the group's terrorism is repeated, when another soft civilian target far away from its Kanuri country base is attacked on Christmas day. 37 people were killed and several others injured in the attack on St. Theresa's Catholic Church in Madalla, just outside Abuja. Abaji based religious teacher (with up to 500 kids reportedly under his tutelage), Kabiru Umar Dikko (aka Kabiru Sokoto) is arrested, tried and convicted for the Madalla terrorist attack.

The arrest of Sokoto, a non-Kanuri, with no links or history to Kanuri Country questioned the narrative that Boko-Haram was largely a Kanuri affair, with some Hausa minority involvement. [80]

[80] https://www.premiumtimesng.com/news/headlines/175497-full-text-video-nsa-sambo-dasukis-famous-presentation-chatham-house-london.html

Top Left - Interior of the UN Building; Top Right - Exterior of St. Theresa's Catholic Church; Bottom- Kabiru (Sokoto) Umar Dikko at trial.

2012 - Former National Security Adviser Owoye Andrew Azazi (a former Army and Defense Chief) died in a naval helicopter crash alongside Patrick Yakowa. Azazi had been recently sacked by Jonathan under immense pressure from 'The North' (euphemism for the Foreign Fuduyan Fulbe factor in Nigeria), While Jonathan suggested that 'consultations with foreign leaders' made him fire Azazi, a leading Newspaper captured seemed more in tune with what most of the public knew. *Traditionally, the North considers the position of NSA as its own. People from the zone had constantly been appointed the NSA until Jonathan summoned the courage to appoint Azazi following the resignation of Aliyu Gusau, who also vied for the presidential ticket of the PDP with Jonathan in 2011. While Jonathan was said to have rebuffed all attempts from the North to recover the seat since he appointed Azazi, the continued bombings and atmosphere of insecurity that* *pervaded the country were said to have made him to change his mind.* [81]

Azazi was replaced by Sambo Dasuki, IBB's first ADC while he was Military ruler, and also the son of the former top Fulbe clan leader in Nigeria.

Kaduna Governor, Patrick Yakowa, and former National Security Adviser, Patrick Azazi died in a helicopter crash.

2014 - Boko Haram gained international attention with the kidnapping of hundreds of schoolgirls from Chibok, a minority village on the fringes of Kanuri Country in Borno State. Often ignored or underreported was the fact that this terrorist group had also killed dozens of schoolboys in Yobe state, shortly before the kidnap of the girls in Chibok. Goodluck Jonathan seemed to assume an unnecessary catch-22 stance of 'damned-if-you-do-damned-if-you-don't'; And for a while, he didn't, as the government was slow to react or take action on the mass killings (in Yobe) and Kidnappings (in Chibok) of secondary school children.

Part of Jonathan's challenge was the sharp criticism he faced when attempting to ramp up action on Boko Haram. Most of these came from 'Northern' (i.e. mostly Hausaland's Fulbe elite)

81 https://punchng.com/news/north-reps-forced-jonathan-to-sack-azazi-defence-minister-investigation/

leaders including serial candidate Buhari who reasoned that 'attacks on Boko Haram were attacks on the North'[82] Goodluck Jonathan seemed overwhelmed, and sometimes unsure and indecisive. A big reason for this was the level of support he received from his office for most of his administration. However, as the leader he bears ultimate responsibility for not identifying weaknesses quickly and making necessary changes.

Chibok girls in the hands of Boko Haram terrorists.

Some have suggested that the 'Niger-Delta' has blown its chance at 'leadership' since Jonathan did not represent the 'Zone' well. This is hilarious, if not outrightly ridiculous, for a number of reasons:

- The Niger-Delta has several capable leaders and achievers, but like most 'normal and responsible' Nigerians,

[82] https://thenationonlineng.net/buhari-faults-clampdown-on-boko-haram-members/

they stay away from the criminal reality show and minefield of Nigerian Politics.

- If the ethnic groups and countries in the region decide to band together and demand the permanent Presidency of a Nigerian state that has been largely funded and sustained by the oil fortuitously located on their lands and waterways, it is worthy of consideration and negotiation, especially since there has been nothing fortuitous about the environmental damage suffered by the region due to the negligence of forces and people that are mostly external to the region. It is up to the other funders of the Nigerian State (Customs Duty, Corporate Income and Value Added Tax providers) to make their demands on the tune they want played for paying the piper.
- Goodluck Jonathan was not chosen by the 'Niger-Delta', as Obasanjo, Ibori and Saraki, and a few others who handpicked Jonathan as Yar'Adua's running mate were not representing the region with their actions.
- Corruption is definitely a cancer, and cases of corruption under Jonathan or any other leader will not be justified or excused. However, the sensational cases of corruption reported under Jonathan, headlined by the billions of dollars of missing revenue calculated from the gap between expected revenue from crude sales and remitted revenue by NNPC have continued in the administration of his successor without a whimper from the Central Bankers, past or current (as at 2021). Specific cases of individual corruption that have been reported are also just as bad, if not worse, given the impact and outcomes (crazy government borrowing, accelerated exchange rate depreciation etc).

- It is difficult for a device to act differently from what it was created to perform and it is virtually impossible to construct a certain type of building when the foundation was designed for another. Nigeria was conceived in corruption, exists for corruption, and operates by corruption. As such the end result and outcomes can only be corruption. It should be noted that the planners, executioners and beneficiaries of the corrupt Nigerian state are almost exclusively foreigners with no longstanding cultural and historical ties to the indigenous peoples and countries that are wholly or partly trapped in the Nigerian State. This is also why this book will waste little time with corruption taxonomies and catalogues.

Vanguard

...towards a better life for the people

ONLINE | www.vanguardngr.com

VOL. 25 NO. 62023 WEDNESDAY, NOVEMBER 27, 2013 N150

HOW THEY STAND	PDP	APC
GOVS	18	16
SENATE	58	48
REPS	147	199

FG overspends by N110bn as 3rd quarter revenue falls — CBN »35

Amaechi, 4 other PDP govs, nPDP join APC

- Kwankwaso, Nyako, Wamakko, Ahmed also defect
- Fear of leadership change grips N-Assembly
- I can't leave the house I built — LAMIDO
- I'm not moving, until... — BABANGIDA ALIYU
- **More stories, reactions** on Pages 54 &55

COLUMNISTS:

AKINTOLA OMIGBODUN •P.24

Mr & Mrs

Asari Dokubo arrested in Benin Republic »9

Frontpage of leading Newspaper 'Vanguard' announced the defection of Amaechi and the newPDP Governors to APC

2015 - Unresolved security challenges and concerns over unchecked corruption contributed to the defeat of Goodluck Jonathan in the 2015 election. Interestingly, these issues have arguably increased under the Buhari administration. What really did Jonathan in however, was the cannibalization of his own (potential, if not active) electoral asset base, which almost directly led to a strengthening of the opposition. Former Lagos Governor, Tinubu had orchestrated a merger of "strange bedfellows" between his ACN and Buhari's CPC, but even that was unlikely to knock off the ruling party without the New PDP (5 Governors that defected from the ruling party to the new coalition), Saraki, and crucially Amaechi, who served as the Director-General of both of Buhari's successful runs at Nigeria's Presidency.

Amaechi, Tinubu and Buhari

2015 - Buhari quickly appoints a GMD for the National Oil company, NNPC, but waits for almost 6 months after his inauguration before announcing a cabinet, famously saying that Ministers are 'Noisemakers'[83]. This must have sounded like a lottery cashout for leading civil servants, for obvious reasons.

2015 - Buhari personally retains Petroleum Ministry portfolio[84] and his faction of the winning 'coalition' naturally gets the lion share of his appointments, to the boundary of obvious

---

83 https://www.pulse.ng/news/politics/buhari-ministers-are-just-noisemakers-like-politicians-president-says/jv3zck3

84 https://www.theguardian.com/world/2015/sep/30/nigerian-president-muhammadu-buhari-to-become-oil-minister-in-own-cabinet

nepotism[85], according to some[86]. However coalition members, and rumoured election sponsors, Tinubu and Amaechi, seem to be well 'taken care of', with Tinubu's camp providing Buhari's Vice President, Yemi Osinbajo amongst other positions, and Amaechi personally taking the portfolios of Transport and Aviation.

Saraki seems out in the cold, but not for long, as he orchestrates a 'coup' in the Senate, and becomes Senate President, with the help of opposition PDP Senators. A similar move in the 2nd arm of the legislature, the House of Representative sees Yakubu Dogara emerge as speaker. Both moves secure victories over the ruling party candidates for the positions.

Dogara with Saraki

2015 - A convoy of the Army Chief, Lt. Gen Tukur Buratai is held up in a Zaria neighbourhood by Shiite (Shia, also sometimes

85 https://dailypost.ng/2017/10/30/81-100-appointees-president-buhari-northern-region-see-full-list/

86 https://www.premiumtimesng.com/news/top-news/189383-buhari-sued-over-lopsided-appointments.html

referred to as Islamic Movement of Nigeria (IMN)) Youth. The Youth were supposedly protesting an earlier incident where members of the group were reportedly attacked by soldiers. Shiite - Army Clashes had occurred over the past year, following a clash during the group's procession that reportedly resulted in the death of the sons of the leader, Ibrahim Zakzaky.

A couple of senior officers in the convoy got down to reason with the youth, many of whom were armed with stones, sticks from the images seen from the incident. Machetes/Cutlasses were also reported. Over the next 24 hours, members of the group mobilised to the compound of the leader to prevent his arrest, which was suspected, but not announced. Various reliable accounts report that almost 350 members of the group were killed as Zakzaky and his wife, Zeenah were injured and arrested.[87]

Senior Army Officers in the Chief of Army Staff's convoy try to appeal to and negotiate with IMN Youth for 'passage rights'

[87] https://www.theguardian.com/world/2015/dec/16/nigerian-army-killings-of-shia-muslims-to-be-investigated

Zakzaky, as an Economics major and students' religious leader/activist in the 70's was impressed by the Khomeini revolution in Iran, and soon started a Shia movement in Northern Nigeria, after spending some time in Iran.

The IMN soon became a noticeable even if small minority group, and even though the group has an objective to create an Islamic State in Nigeria, it is not known to have taken up an armed struggle against existing religious, traditional and political institutions, or kill indigenous peoples and take their lands. This is a contrast from the Foreign Fuduyan Fulbe takeover of Hausaland or the dozens of similarly minded individuals and groups that have attempted to introduce their own brand of religious regime since.

Quite naturally, the growth of the IMN led to a proxy war between Iran and Saudi Arabia, and Senior Saudi Royals 'reportedly boasted of Riyadh's success in the suppression of El-Zakzaky'[88] with respect to the Zaria incident and the Group's travails, according to IMN Spokesman Aḅdullahi Mohammed.

While a proxy war between Iran and Saudi Arabia is not unexpected, there is an interesting and very local dimension to Zakzaky's troubles with the Fulbe rulers of Hausaland. The Fuduyan takeover of Hausaland was by foreigners, who identify and can easily be identified as such. The takeover has remained effective for two centuries as the foreign clans maintained a stronghold over the traditional, religious and political institutions and symbols of authority in Hausa Country.

Zakzaky represents a small (as at 2021) but genuine threat to that stronghold. An indigenous and alternative symbol of

---

88 https://punchng.com/el-zakzaky-1886-children-orphaned-200-arrested-since-zaria-massacre-says-imn/

religious authority. He is not claiming to be Arab or Saudi, just a Hausa religious leader.

While an overwhelming majority of Hausas are still subject to the Wahabbi aligned Fulbe religious authorities, the thought of an alternative, fully indigenous religious authority and institution in Hausaland is extremely discomforting to the ruling clans.

It was definitely not surprising that there would be a major incident as soon as Buhari became Nigeria's President. Buhari had always seemed eager to prove his 'Fulbeness' to the ruling clans, given his mother's Kanuri background and his somewhat lower origins in their ranking system. He was reportedly born in a camp of migrating Fulbes along the current Niger - Nigerian border, and many have used this to highlight that he has no cultural, ethnic and historical ties to any indigenous Nigerian groups. The Fulbe 'cabal' supposedly led by Buhari's older nephew Mamman Daura (who lives in the Presidential Villa complex) seemed to have milked this 'desire to belong' extensively.

A strange twist to this, is that Buhari also sees himself as some sort of living armistice to the centuries' long war between the Fuduyan Fulbes and the Kanuris. Fuduye and Mohammed Bello, having killed off or deposed the rulers of Hausaland, turned their attention to the Kanuris (Kanem-Borno), Hausaland's eastern neighbors. El-Kanemi, the Kanuri leader, was better prepared than the Hausa rulers though, as the Kanuris have had to move base and capital over time due to various armed adventurists with religious claims trying to take over their country before and after the Fuduyans. Most famous among these was the Sudanese Rabih-az-Zubayr.

Buhari seemed to pull both groups (one foreign, the other, only partially on Nigeria's North Eastern Fringes[89]) into a strong ruling alliance that goes far beyond a kitchen cabinet[90]. That, sadly, is the reality of the Nigerian State cobbled together by foreigners and run for the benefit of foreigners. One where the capacity of indigenous peoples and countries to function, cater to their welfare, and defend themselves has been taken away. One where foreign militia and terrorists roam the country seemingly unchallenged, sacking villages and taking land[91], while attempts

are suppressed by a Military that amazingly somehow uses the insignia that the Foreign Fuduyan Fulbes used while taking over Hausa country.

2018 - Alex Badeh, former Chief of Air Staff and Chief of Defence Staff is gunned down in the suburbs of Abuja despite the presence of his security team.

2018 - With Buhari conveniently out of the country and power "legally transmitted" to his Vice President, the long awaited move against Saraki happens (there had been several 'minor' EFCC arraignments, and a ridiculous link to a major armed robbery in Offa) with the secret police (DSS) sealing off the National Assembly for reasons that officially remain unclear[92]. The move, which immediately draws widespread condemnation,

---

89 Long term Chadian strongman, Idris Deby was killed in a battle with CACT rebels in the 'Kanem' region of Chad. A spokesman for the FACT rebels had the same surname as Nigeria's special envoy to Chad on the crisis. The envoy is a long term Nigerian power broker of Kanuri origins.

90 https://www.vanguardngr.com/2019/05/kanuris-influence-on-buhari/

91 https://dailypost.ng/2021/07/27/fgs-silence-on-insecurity-a-continuation-of-1804-jihadist-movement-ortom/

92 https://www.reuters.com/article/us-nigeria-politics-idUSKBN1KS0R1

is reportedly as much of a surprise to the public as it is to the Vice President, Yemi Osinbajo, who immediately summoned Lawal Daura, head of the agency.

Daura reportedly condescendingly reminds Osinbajo of his political irrelevance in the scheme of things, and Osinbajo, said to be unable to reach Buhari announces the suspension of Daura, and replaces him with Matthew Seiyefa, in an acting capacity.

The move by Osinbajo, his most aggressive as Vice President, was met with an almost instant repercussion. Şeiyefa was promptly replaced with Yusuf Bichi[93], and Finance Minister, Kemi Adeosun, one of Osinbajo's two Ministerial appointees in his role as coordinator of the economy, resigned under circumstances pertaining to her exemption certificate from the National Youth Service. The other 'Osinbajo' Minister, Okey Enelamah, a Private Equity executive in charge of the Trade, Investment and Industry Portfolio is not retained, as Buhari is re-elected in 2019.

R-L Buhari, Saraki and former Central Bank Governor Lamido Sanusi

[93] https://guardian.ng/news/outrage-as-buhari-removes-dss-chief/

2019 - Former Babangida ADC, Jonathan's National Security Adviser and son of former Fuduyan Fulbe clanleader, Sambo Dasuki, in detention despite several prior court orders granting him bail, is released alongside activist and Sahara Reporter's Publisher, Omoyele Sowore. Dasuki is alleged to have 'shared' $2billion meant for Arms procurement to various political associates,[94]but his troubles may have been compounded due to Buhari's long-standing grudge over his (Dasuki) role in the IBB coup.

Sowore's release seems to have been due to some US pressure[95], with Dasuki thrown into the mix to 'equalize' things, as is common with the ruling clans. However, still illegally detained by Buhari was Shiite leader, Ibrahim El-Zakzaky, who is seen as a risk to Nigeria's ruling Fulbe class that controls the religious, traditional, and government institutions in Hausa land.

Zakzaky was released in 2021, as the Buhari Government prepared for the (separate) trials of IPOB leader Nnamdi Kanu and Sunday Igboho who escalated his protests of Fulbe herdsmen killings to rallies and demands for a separate Afeire Nation. This release of Zakzaky suggests an attempt to ensure there is no lingering and potentially game-altering opposition to Nigeria's Fuduyan Fulbe rulers from their key platform (or subjugated country) of Hausaland.

94 https://www.bbc.co.uk/news/world-africa-34973872

95 https://www.premiumtimesng.com/news/headlines/369206-u-s-lawmakers-write-nigerias-agf-malami-over-sowore.html

Dasuki, El-Zakzaky and Sowore

2019 - In a brazen, unlawful and almost unprecedented move[96], Buhari removes the Chief Justice of the Country, Walter Onnoghen, in the build up to the 2019 elections.[97] The move is seen as ensuring that there would be no surprise through the courts, as Buhari prepared to square up against Atiku Abubakar in the elections. Buhari termed the removal as a suspension to

---

[96] *Murtala removed Justice Teslim Elias from as Chief Justice following the coup against Gowon in 1975*

[97] https://www.economist.com/middle-east-and-africa/2019/02/02/nigerias-president-sacks-the-chief-justice-weeks-before-an-election

bypass the requirement of getting two-thirds of the senate of sanctioning the move.

2019 - Unsurprisingly, Muhamadu Buhari, whose rule as Military dictator left the Nigerian economy in shambles, and who did not even seem to be interested in attempting to fulfill his electoral promises during his first term as civilian ruler[98], is re-elected as President, a development that some discerning minds believe to be the countdown to the death knell of the clearly failing Nigerian State[99]. Buhari, is, however, extremely popular in the far north of the country even though his image had been carefully managed by an extensive media network, starting from his older nephew, Mamman Daura, who was editor of the conservative New Nigerian newspaper between 1969 and 1975. His inner circle also include Isa Funtua, a past President of the Newspapers Proprietors Association of Nigeria, and Abba Kyari, who worked with Daura at New Nigerian, and was Buhari's first Chief of Staff. It was no surprise that an organisation with scores of staff called the Buhari Media Organisation, was set up to launder Buhari's image, and counter any stories, true or not, considered not to favour the public image that had been built up.

2021 - Former military officer, doctor and self-appointed negotiator and spokesman of terrorists, Ahmad Abubakar Gumi (Jr), blatantly lied in a meeting with the terrorists that non-Moslem soldiers were attacking moslem communities, attempting to create a division and break the camaraderie of officers serving in

---

[98] https://www.thecable.ng/promoted-behold-the-62-failed-2015-election-campaign-promises-of-muhammadu-buhari-and-apc

[99] https://www.vanguardngr.com/2021/07/nigeria-witnessing-the-epilogue/

the military. Nothing mentioned about the high school student killed in cold blood as the terrorists kidnapped the victims.

The Government appeared to back Gumi in statements that seemed incomprehensible to sane, rational minds. False rumours of Gumi Jr's invitation by the military and intelligence agencies were later denied[100].

A few days after Gumi's remarks and meeting with the terrorists in Niger state, a Beechcraft 3501 surveillance aircraft that was involved in the search for the kidnapped Niger state schoolboys, with a seven-man crew, surprisingly all non-Moslem, in a very religiously diverse Nigerian Military, crashed shortly after a refueling stop at the Abuja airport. While the official report of the crash has not been released, foul play or sabotage has not been ruled out, given Gumi Jr 's comments, the timing of the crash, and the identity of the victims.

100 https://www.vanguardngr.com/2021/06/i-was-not-invited-arrested-quizzed-by-dss-sheikh-gumi/#:~:text=Sheikh%20Ahmad%20Gumi%2C%20the%20Kaduna,allegedly%20made%20against%20the%20military.

Images from an official Military announcement of the transition of the Officers and Men killed on a mission trying to locate the position of the terrorists that kidnapped Students in Niger State in February 2021. Perhaps an invitation of Gumi Jr by even one of several budget consuming agencies (Police, DSS, DMI, NSA, NIA etc) would have confirmed the location of the terrorists, making the surveillance flight unnecessary.

2021 - Chief of Army Staff, Lt. Gen Ibrahim Attahiru, and 10 other officers and men of the military died in another Beechcraft 3501 Aircraft crash, this time, as the plane attempted to land in bad weather, and was redirected to the civilian Airport in Kaduna[101]. The Army Chief had only been in office for a few months, and was going to attend a military graduation ceremony.

Public information about military crash inquiries are hard to come by in Nigeria, so additional details regarding the cause of the crash may be unknown for some time, outside of 'inclement

[101] https://www.legit.ng/1416930-list-names-10-officers-died-alongside-general-attahiru-plane-crash.html

weather conditions' at the time, which may, or may not have been a major determinant.

2021 - President Buhari departs the country for London in late July for a postponed[102] medical trip that now includes an Education donors conference. Bola Tinubu is curiously in London at the same time, so a 'state of the alliance talks' may also be on the unannounced agenda. Despite vowing to oppose medical tourism if elected as President during his campaign[103], Buhari had spent over 200 days on medical trips[104] to London as at 2021, with his wife, Aisha, suggesting that the Buhari Government did not have the competence to ensure the Presidential Villa Clinic was properly run and adequately stocked with the most basic medication including syringes[105].

---

102 https://www.premiumtimesng.com/news/top-news/469999-buhari-postpones-london-medical-trip.html

103 https://www.theguardian.com/world/2016/jun/07/muhammadu-buhari-nigeria-ear-infection-medical-treatment-uk

104 https://www.premiumtimesng.com/news/headlines/477336-timeline-buhari-has-spent-200-days-in-uk-for-treatment-since-assuming-office.html

105 https://www.vanguardngr.com/2017/10/urgent-aisha-buhari-calls-for-probe-of-state-house-medical-centre-over-poor-facilities/

London has been a choice meeting point for Buhari and Tinubu over the years.

2021 - Terrorist attacks and kidnappings continue in many parts of the country, with Southern Kaduna and Plateau remaining among the key flashpoints. In Kaduna, 13 schools, mostly run by religious organisations are closed following the kidnapping of scores of students from Baptist High School[106]. In the neighbouring Jos Plateau indigenes accuse the GOC of one of the Army's major divisions of being on an "evil mission to ... wipe out the chiefdom ' since the marauders attack in broad daylight, steal, kill, rape and destroy farmlands."[107] Those fears followed complaints about what seemed to be a pattern of the Army not intervening when foreign Fulbes (based on eyewitness accounts)

106 https://punchng.com/breaking-abduction-kaduna-orders-closure-of-13-schools/?utm_medium=Social&utm_source=Twitter&amp&utm_medium=Social&utm_source=Twitter&__twitter_impression=true

107 https://dailypost.ng/2021/08/10/plateau-massacre-buhari-orders-military-to-kill-illegal-weapon-holders/

attack communities with high grade weapons, but eagerly intervene when communities rally to go after the killers.

Many wonder why the Nigerian Army still retains the insignia from the flag of an invading foreign Fulbe militia, as it took over Hausa Country, and if this this signals the ideological and spiritual purpose of the Nigerian Army, and the Nation it helped forge, and has struggled (rightly or wrongly) to keep together.

A cartoon in the Punch, a leading newspaper, captures the widely held thought by indigenous Nigerians on the stance of the Military regarding invading foreign terrorists.

In what seemed to be an exclamation mark on the security situation (even if calling it that is diversionary), two commissioned officers were killed by terrorists, and a third kidnapped from the Officers Quarters of the Nigerian Defence Academy (think of West Point, USAFA and the Naval Academy all rolled into one). In what seemed to be a symbolic insult on the entire Military, the Officers were from the three Armed Forces with Major Christopher Datong (Army) abducted, and Lt. Cdr Wulah (Navy) and Flt Lt Okoronkwo (Air Force) killed.

On the same night (or the next) as the NDA attack, a four-hour drive to the Southeast, terrorists killed over 30 indigenes in their homes[108], following widely expected attacks that had seen evacuations of non-Indigenes from the University of Jos.[109]

In the Jos attacks, and in previous attacks around the NDA vicinity[110], foreign Fulbe Militia, or Bandits, as the Herdsmen without cattle are now called, were confirmed by victims[111] to be their attackers and abductors[112].

Nigeria does not have a security problem per se. It has a structural problem, and an invasion problem. This can be resolved easily by the constituent countries picking up the gauntlet and functioning as countries to protect life and property.

The military and the police cannot even protect itself from the raving militia of foreign invaders, but are quick to kill indigenes trying to protect their lives and lands. Every country in the Nigerian nation is sustainable at levels far above subsistence. The priority of each country MUST be the protection of the lives, lands and livelihoods of its indigenes and legal residents.

Fulbes are indigenous to at least six nations of their own in West Africa, and face no natural or man made threats in these countries, which are not contiguous to the Nigerian State. Fulbe's are also at the political helm in most of these countries. The Fulbes passing through or residing in Nigeria, especially the ruling clans of Hausaland, know where they are from, and have close ties with their home countries. That is why Fulbe Cleric,

---

108 https://www.vanguardngr.com/2021/08/breaking-gunmen-invade-jos-north-community-in-fresh-attack-over-30-killed/

109 https://independent.ng/plateau-killings-nans-group-evacuate-279-fct-students-from-jos/

110 https://punchng.com/parents-lament-as-kaduna-confirms-abduction-of-39-students/

111 https://punchng.com/breaking-kaduna-forestry-college-students-released-56-days-after/

112 https://www.channelstv.com/2021/04/27/abducted-pregnant-student-others-call-for-help-in-new-video-released-by-bandits/

Dahiru Bauchi will, from his people, marry the daughter of *Senegalese* Cleric, Ibrahim Inyass. Similarly, Aisha Buhari's Chief Guest at her son's Wedding activities will be wife of *Gambian* Fulbe President Adama Barrow.

Perhaps the only difference is that the Fulbe ruling clans of Hausaland believe that they must maintain an extremist and intolerant religious stance is for a public show and claim to credibility. For instance, Umar Tall stayed with Mohammed Bello in Sokoto for years, to gain practical lessons in 'how to invade and destroy previously sustainable societies', although his attempts had mixed results in their homelands). On a related note, the clans naturally believe in using the wealth and influence of the 'Nigerian Countries' to exert influence and achieve their objectives in their home Countries.

Aisha Buhari and her Fulbe Guests from Home.

The lives of the millions of indigenous Nigerians cannot be mortgaged to ruling rights of the Fulbe ruling clans of Hausa Country, and the plundering and killing 'rights' of their affiliated Bandits, Herdsmen, Militia, or whatever other label with which the terrorists are tagged.

STATE OF THE (DECOLONISED?) NIGERIAN COUNTRIES IN 2021

The economy has significantly worsened under Buhari as have security matters (either by design or happenstance). The Boko Haram insurgency now seems to have settled to an equilibrium of ISWAP occupying territory in the Northeast of Nigeria following the defeat of its rival factions in the group[113]. ISWAP leader Al-Barnawi announced the demise of his former comrade in terror turned rival, Abubakar Shekau[114] as the groups battled for territory and supremacy, with an obviously overstretched military almost reduced to the role of a distant observer.

A new flank of terror is in full swing on the Northwestern front of the country where terrorists, now officially nicknamed bandits[115] by the government, started their activities in the state of Zamfara, and have slowly worked their way down to Nupe Country and further down to the southwest. Some of these terrorists, mostly Foreign Fulbe's, were according to politician, Kawu Baraje, invited to the country to help overthrow the government of Goodluck Jonathan if Buhari did not win the 2015

---

[113] https://www.reuters.com/world/africa/boko-haram-cleric-confirms-shekaus-death-urges-fighters-loyalty-2021-06-18/

[114] https://www.reuters.com/world/islamic-state-west-african-province-says-nigerias-boko-haram-leader-is-dead-2021-06-06/

[115] https://www.brookings.edu/blog/africa-in-focus/2021/02/18/rising-insecurity-in-northwest-nigeria-terrorism-thinly-disguised-as-banditry/

election.[116] Unfortunately, they are now in hundreds of camps in the forest reserves and it seems the military is unable to return these terrorists to their countries of origin.

Gumi Jr., whose father and Ahmadu Bello started JNI, perhaps to whittle the authority for the (Nigerian based)[117] recognized Fuduyan Fulbe clan leader and Bello rival in the 60's[118], has stated that there are over 100,000 of these men[119], armed and ready to cause mayhem in Nigeria (or according to some, just continue the foreign violent propagation that started in Hausa Country). Gumi. Jr, who has assumed the role of spokesman and negotiator of these terrorists, demanded amnesty and payments[120], and also announced new strategies and intended actions of the foreign terrorists, but his statements were downplayed by the Military. Notable among these, was his statement that the terrorists on the Northwestern flank were in the process of getting anti-aircraft missiles[121], a threat or risk that quickly crystallized in the following months as the terrorists downed a fighter jet piloted by Flight Lieutenant Abayami Dairo[122]. Dairo

---

116 https://barristerng.com/kawu-we-brought-in-fulani-from-mali-but-after-election-they-refused-to-go-back/

117 There are competing claims flowing back to Lugard's deposition of Attahiru that the legitimate head of the Fulbe clans is (in) Mai Wurno, on the bank of the blue Nile in Sudan based on the instruction Fuduye said he received and passed on.

118 The older Gumi criticized the Fulbe clanleader for accepting the British Honours, but as sidekick to Ahmadu Bello softened up to the colonials, and had several sons and close relatives serve in the Nigerian Military, including Gumi Jr. He died in a London Hospital in 1992

119 https://hallmarknews.com/there-are-about-100000-bandits-in-north-west-alone-no-school-is-safe-until-gumi/

120 https://www.bbc.co.uk/news/world-africa-57007326

121 https://www.thecable.ng/bandits-gathering-money-to-purchase-anti-aircraft-missiles-says-gumi

122 https://www.bbc.co.uk/news/world-africa-57893662

ejected and was rescued by an Army unit. This outcome was a lot more positive than the situation with Wing Commander Chimda Hedima, whose jet was downed by Boko Haram, but was found by the terrorists. Hedima, born on October 1 (Nigeria's independence day), son of a retired Air Force Officer, whose homeland is in the region of the Chibok area terrorized by Boko Haram, was brutally murdered on Camera.

L: Wg. Cdr Chimda Hedima R: Flt. Lt Dairo received by Chief of Air Staff, Air Marshall Amao, and Chief of Defence Staff, General Irabor and other senior Military Officers

Boko Haram and ISWAP on Nigeria's northeast are believed to have had anti-aircraft weaponry for a number of years, and the scaled-up capacity of the terrorists on the northwestern front is due to collaboration[123] and cooperation between the two terrorists groups[124]

Completing the trifecta of terror converging in (or from) Nigeria's Northern fringes is the Foreign Fulbe herdsmen, on

[123] https://www.france24.com/en/live-news/20201216-nigerian-gangs-kidnapped-children-on-behalf-of-boko-haram-sources

[124] https://issafrica.org/iss-today/boko-haram-teams-up-with-bandits-in-nigeria

its own identified as the fourth most dangerous terrorist group in the world.

Sadly, groups like Boko Haram, uprisings like Maitatsine, and even the now-nicknamed Bandits show the danger of religious extremism and terrorism which usually starts when a couple of men get together and decide that their own form of religion is more puritan than the existing one or other peoples'. This is no problem on its own. There are many puritan religious societies that withdraw from society to form puritan communities like Aiyetoro in Nigeria's Afeire country and various Amish and Mormon groups in the USA, with the occasional but limited government oversight to prevent the incidence of exploitative cults.

The problem arises when these puritans, extremists, or outright criminals with a religious cover, choose not to practice their religion peacefully and live (economically) productive lives in the context of a sustainable society, but embrace violent propagation as a means to terrorize people, disrupt society and ultimately get paid (as demanded through taxes, ransom, or from eventual rulership of the cities or societies) for their value and society destroying propagation.

They obviously have no intention of working with their hands and minds to build productive or sustainable societies, so rather than withdrawing to form their own puritan societies to worship as they please or deem right, they set their sights on using terror and violence to steal, kill, destroy and systematically take over existing societies in which they are mostly aliens who have enjoyed the warm welcome and hospitality of the indigenes.

Consequently, the three aggressive terrorist groups, with mostly foreign ideologies, origins [125]and leadership have enjoyed tacit government support and even collaboration[126] in a Nigerian State (which is itself a collaboration of foreigners between the Fuduyan Fulbe and English Colonials) under a leader whose entrance and pathway through public life has been one of nepotism, discarding of standards, threats, illegal use of violence and coups.

It is therefore not surprising, and should even be expected, that while the government has seemingly accepted the inevitability of these aggressive terrorist groups on the country's Northern fringes, with complaints of an overwhelmed and overstretched military[127], reactions to the invasion, ethnic cleansing, kidnappings and land grabbing activities of these foreign groups (mostly foreign Fulbe militia claiming a Fuduyan right and ownership to all of Nigeria), are often instantly crushed by the Nigerian secret police and the military[128]. The Nigerian Government also hilariously (actually, seriously, by its own standards) blames the invading terrorists on its Northern fringes and their high calibre weapons on the Western overthrow of Ghadaffi in Libya[129]. Ghadaffi, of course had his thoughts on the Nigerian State[130],

---

125 Boko Haram's Kanuri origins are in Maiduguri, but its operating base and interests are the Lake Chad area which spreads across at least four countries.

126 https://punchng.com/bandits-ready-to-listen-believe-sheik-gumi-than-govt-lai-mohammed/

127 https://www.vanguardngr.com/2020/12/military-overwhelmed-by-security-challenges-govs/

128 https://punchng.com/buhari-and-the-three-musketeers/

129 https://www.premiumtimesng.com/news/top-news/264764-buhari-blames-gaddafi-for-killings-across-nigeria.html

130 https://www.reuters.com/article/ozatp-nigeria-gaddafi-20100329-idAFJOE62S0CD20100329

having visited during what was (wrongly) previously believed to be its lowest possible low.[131]

Despite the restructuring of the Nigerian state being an anchor of his Presidential campaign in 2015[132], and his party setting up an internal body (headed by Fulbe Kaduna Governor and former FCT Minister and head of the country's privatisation agency, Nasir El-Rufai) to codify the ruling party's position on restructuring [133]during his first term, Buhari has suggested not knowing what the term means, and has waived off long standing foundational and structural problems of the Nigerian state[134], as the language and stance of his spokesmen, representative and acolytes wave on invading foreign terrorists in the subjugation of indigenous Nigerians[135].

To be fair to Buhari, he has not claimed to be studious or bright[136], famously failing to show proof of the basic school leaving certificate required to contest the Presidential elections, and possibly not needing it to join the Army due to the Bello-Ribadu policy. He has also publicly stated that his peer group in the military spent time planning coups instead of self and career development. It was no surprise that Buhari rarely gave 'off the cuff' speeches or interviews that lasted up to 3 minutes during his well managed 2015 campaign, and most enduring memories

---

131 http://www.aparchive.com/metadata/youtube/f08f2a63e64fd0921efe69cb65bdc2a2

132 https://punchng.com/nigerians-await-apcs-promised-restructuring-five-years-after/

133 https://www.thecable.ng/state-police-fiscal-federalism-12-recommendations-made-by-apc-committee-on-restructuring

134 https://newspeakonline.com/oyegun-laments-apcs-failure-to-fulfill-promise-on-restructuring/

135 https://thenationonlineng.net/why-is-president-buhari-afraid-of-restructuring-i-think-i-know/

136 https://www.youtube.com/watch?v=DB__J5YE4d0

of the ruling party campaign rallies, have been the dance steps of the key party officers (absolutely nothing wrong with dancing, which is a key part of all indigenous cultures in Nigeria).

Address the issues? Nah, let's just dance.

Despite the clear seismic-type fault-lines facing the Nigerian state on multiple fronts, political activity is already in full swing as at mid-2021 as various politicians battle to succeed Buhari and gain control of a clearly failing Nigerian State, which has outlived the purpose its creators needed it for, but still has some value in the ability of its rulers to control \$15 billion - \$35 billion of oil revenues (a large part of that off the records through the National Oil company) annually.

SUCCEEDING BUHARI

Starting with the ruling party, which is seriously fractured as at 2021, there are at least four identifiable groups and possible factions vying for Nigeria's Presidency.

Leading this group is Buhari's 'managing partner' of the rul-

biggest basis/claim for the Presidency seems to be a 'payback agreement' for providing the key missing pieces for Buhari's quests for the Presidency in 2015. Tinubu, had been increasingly shut out from party affairs, with not so subtle signs ranging from the removal of the party Chairman that was deemed friendly to his ambition, former Edo Governor, Adams Oshiomole, to Buhari's Freudian slip in one of his few live interviews that '*"You cannot sit there in Lagos, for instance, and decide on the fate of APC on zoning," "Therefore we should allow the party to decide."*[137]

This would normally be enough to end a politician's prospects, given the influence of the office of Nigeria's President in political and local affairs, but both men curiously timed a visit to London at the same time a few months later, and keen observers of Nigeria know that London is a favorite destination for crucial talks and negotiations on the fate of Nigeria (with Sudan and Saudi Arabia reportedly being other interesting locations).

Femi Gbajabiamila gets endorsed by Fulbe cleric Dahiru 'Bauchi' in Saudi Arabia for Speaker of Nigeria's House of Representatives.

[137] https://guardian.ng/news/you-cant-sit-in-lagos-and-decide-the-fate-of-apc-on-zoning-buhari/

In all of this, it is unlikely that due (if any) consideration will be given for the fate of long-suffering indigenous Nigerians, millions of whom now lie below a poverty line that binds them together as the largest collection of poor people in the world.

The group that currently seems to control the ruling party (as at 2021), is in the image and likeness of Buhari, who views himself as the armistice between the Kanuri and the Fulbe rulers of Hausaland. This is also why he had a soft spot for Sani Abacha,

coup in 1983.

Fulbe Governor of Kebbi and Abacha lawyer Abubakar

the late dictator[138], and is reportedly in line[139] for a $110 million[140] settlement is pulling the strings, while Kanuri Governor of Yobe, Mai Mala Buni, recently married to one of Abacha's daughters, has forced through a couple of court suits and orders (obeying some and defying others), to emerge and remain as the 'Caretaker Committee' Chairman of the party.

This group is believed to have Buhari's strong backing, and is working for his interest, whatever that may be.

Chibuike Rotimi Amaechi served as Campaign Chairman of both of Buhari's successful Presidential runs, and was a key part of the coalition that got Buhari elected in 2015, and was rewarded with full control of the country's transportation system (Sea, Land and Air). He has maintained a stoic silence on questions regarding a Presidential run, often deflecting by saying he

---

138 https://www.justice.gov/opa/pr/us-forfeits-over-480-million-stolen-former-nigerian-dictator-largest-forfeiture-ever-obtained

139 https://www.vanguardngr.com/2020/03/revealed-why-us-is-stalling-repatriation-of-308-abacha-loot/

140 https://www.thetimes.co.uk/article/dictators-henchman-in-line-for-87m-payout-hjlvct9tc

is focused on delivering rail infrastructure to the country. His influence, like that of Tinubu, has expectedly waned in Buhari's second term, with the Aviation Portfolio he previously headed handed over to Buhari's Sister's son, Sirika Hadi.

Amaechi's loss of control of the political machinery of his home state (to Nyesom Wike) has also been brought up, but this is not much of a problem if the various 'power brokers' decide to go along with him, as was the case with Obasanjo in 1999, who became President, fresh out of Prison, with his entire region firmly opposed to his candidacy.

Some political watchers however view Amaechi as a trusted ally that Buhari can use to quell the growing disillusion of the Ndigbo with the Nigerian state, and he recently (2021) asserted his authority with the sudden firing of the MD of the Nigerian

the key transport agency, was seen as part of the operating and governance arrangement of the Buhari centered ruling coalition. She was a BBOG Co-founder, and previously Chief of Staff to Kaduna Governor, Nasir El-Rufai (who alongside Amaechi and Saraki defected from Jonathan's PDP in the buildup to the 2015 elections).

L: El-Rufai and Amaechi driving Jonathan out of power! (It was a drive to the DSS Office), R: Bala-Usman acknowledged as her 'Boss Forever' after Amaechi fired her.

Rounding up the list of people with declared or rumored ambition in the ruling party are former Governors and serving Senators Orji Uzor Kalu and Rochas Okorocha, and current Governors Dave Umahi and Yahaya Bello.

In an interesting twist, moves were made to bring former President, Goodluck Jonathan into the APC by the Buhari (or Bagudu-Buni) faction of the party as the faction sought the best way to preserve its interest beyond 2023. Buhari had spent most of his Presidency blaming the Jonathan administration as the worst and most corrupt in Nigeria's history, but it can always get worse in Nigeria, and it apparently did, judging by Buhari's performance. The Buhari Presidency, had suddenly warmed up to Jonathan in late 2020.

Bagudu, Buni and other APC Governors visit Jonathan

Saraki and the PDP 'reconcilers' visit to beg Jonathan not to leave

Jonathan was quickly begged and convinced to stay in the PDP by the 'reconciliation group' of the party, headed by Bukola Saraki, himself only recently back in the party after his sojourn in the APC.

Apologies if your head is spinning a bit at this time, but there is more to come.

In the PDP, former Vice-President, Atiku Abubakar who returned from a failed attempt at the APC ticket in 2011 to clinch the PDP ticket in 2015 to launch an unsuccessful bid against Buhari (who had removed the Chief Justice a few months earlier to cover any surprises from the Court challenges, which expectedly followed the elections).

Atiku had knocked heads with Rivers Governor (and former Amaechi Chief of Staff turned thorn in flesh) Nyensom Wike over control of the Party, but both had reconciled as at mid-2021 [141]and now seemed determined to remove Party Chairman, Uche Secondus, whom Wike largely installed.

This new Atiku-Wike alliance will be favourite to determine the party's candidate for the 2023 elections, although it will be

---

[141] https://dailypost.ng/2021/07/29/i-have-reconciled-with-gov-wike-atiku/

premature to rule Saraki out, depending on how much of the party he was able to 'reconcile' to himself. Former President Jonathan also remains eligible to run, even though he has been largely inactive on political matters. Others, including former Federal House Speaker and Current Governor, Aminu Tambuwal cannot also be completely ruled out.

A fun fact about Atiku and Saraki is that both have had their claims of Fulbe ethnicity disputed at crunch time by those who should know.

A recent development in Nigerian politics is the symbolic extension of the Bello-Ribadu exemption (or approval) to the political space. In the 60's, young Hausa speaking men of other ethnicities, faked their identities to get a free pass to the officer corps of the military. This practice continued in much lower measure in the Defense Academy era, due to much lower standards required to meet allocated quotas in certain states.

The Fuduyan Fulbe rule of Hausaland and all its institutions, ensures that a couple of foreign clans are able to leverage on their control the largest (partly) indigenous Nigerian country to largely influence the affairs of the Nigerian State.

As such in a weird and twisted way, you have these aspirants for political office make claims about Fulbe ethnicity, which itself should be a disqualifier from any type of office or position in a country seeking any form of development and progress.

Starting with Atiku, his Fulbe claims (which he desperately sought to validate by his ability to speak Fufulde and the fact that he reared cows as a child), was aggressively torn to shreds by terrorist group 'Miyetti', as Atiku squared up with Buhari (or Mamman Daura as some would say) in the 2015 elections. [142]

---

[142] https://punchng.com/that-atiku-speaks-fulfulde-doesnt-make-him-a-fulani-man-miyetti-allah/

Fulbes first came to the area below Kanuri country in Nigeria's Northeast, when adventurist and violent propagator Modibo Adama Hasana 'collected' a flag from Fuduye to go kill and destroy indigenes of the area (who had no business with these Fulbes from over a thousand kilometres away, and had not wronged them in any way) and become 'Ruler of the Southlands'.

To be fair to Saraki, he is not known to have made any puzzling statements regarding ethnicity. That honor would go to his father, fellow medical doctor, Politician and Second Republic Senate Leader, 'Oloye' Olusola Saraki. Perhaps, angling for a way for his rumored Presidential bid to be 'acceptable' to various parts of the country, the Fulbe ethnicity was thrown into the mix.

The Buhari Coup of '83, put paid to the 2nd Republic and the older Saraki's ambition, but it was enough to anger his former friend turned political rival, notable lawyer AbdulGaniyu Folorunso Abdulrazak, to repeatedly attack such claims.[143]Both families have since contended for control of their local Ilorin politics. Ilorin, significantly, is the traditional military cantonment (war camp) of the Alaafin of Oyo.

[143] http://saharareporters.com/2010/12/09/saraki-not-kwara-alhaji-abdulganiyu-folorunsho-abdulrazak

BEYOND POLITICS

Beyond the current (2021) political permutations, the Alaafin had first given permission for the previously unknown Fulbe herders to cross Jebba to occasionally graze cattle at about the time the herdsmen were aiding the Fuduyan takeover of Hausa Country. Like the indeginous rulers of Hausa Country before him, the tendency of the African to welcome and be kind to strangers beclouded the realization that itinerant herders roaming the country unchecked was a perfect cover for reconnaissance and gaining information about the country.

The Alaafin soon lost his War Camp, and eventually his Capital of Oyo to the invading Fulbe clans, with acts of murder, terrorism and raids on defenseless villages by gangs and militia given a religious slant and justification. His fate was better than that of the Hausa Sarkins, who lost their city-states, country and in most cases, their lives under the pretext of a clash between one of them in Gobir, and a preacher that had been retained in the palace for religious duties. The foreign herdsmen, who make ridiculous claims about owning all of Nigeria and more, base these claims on the role herdsmen played during the systematic Fuduyan takeover of Hausaland.

These terrorist raids have continued in spurts for two centuries, with each city that succumbs to foreign Fulbe rulership becoming the Springboard for terrorism on the surrounding cities and regions. As it was with the destruction of a village near Iseyin in 1821, which resulted in the capture of a 12 year old Ajayi (later Crowther), so it was much more recently with Isa Pantami volunteering to lead a squad of killers to Langtang[144] and other villages in Nigeria's Plateau State, and it continues unabated as terrorist militia from far away burn villages[145], rape women and kill in what seems to be senseless, but may be explained as an attempt to create an imported ruling clan lording over a diverse collection of countries through terrorism.

Hausaland appears to be gone, Nupe country seems next in line, Southern Kaduna[146], the Plateau, Tiv country and most of Benue are under constant attack, Ibarapa in Oyo, Sagamu - Ore - Benin expressway, Ibadan - Ife - Ijesha - Ekiti expressway and just about everywhere in Nigeria is under threat from marauding invading terrorists and criminal gangs. There is a worrying escalation of this crisis in Igbo country (Eastern Nigeria) as what started as an effort to chase herdsmen out of forests[147] is now escalating as suspected IPOB Militia or Unknown Gunmen have attacked police stations and other security locations.

---

144 https://www.icirnigeria.org/pantami-as-a-metaphor/

145 https://sundiatapost.com/plateau-killings-we-lost-70-people-in-4-days-to-fulani-bandits-community/

146 https://guardian.ng/news/thousands-flee-zangon-kataf-kauru-as-herdsmen-sustain-massacres/

147 https://dailypost.ng/2021/05/04/igbo-group-warns-killer-herdsmen-to-flee-eastern-forests/

The large cities which seem exempted are all within striking distance. The ruler of the Fulbe clans has admitted that about 70% of the kidnappings in the country are done by members of his clan. Their homeland is not contiguous to any part of the Nigerian state, so there must be a coordinated operation, unifying purpose and obvious weakness bringing these terrorists to the Nigerian State, but not any of its coastal neighbours to the West, which are closer to Fulbe country. No indegeous Nigerian ethnic group or country has militia traveling hundreds and thousands of miles away from its own land and country to kill, rape and plunder.

The Nigerian state was set up by foreigners to plunder the resources of the country and serve the interest of these foreigners over time. The complete devastation of the indigenous countries does not have to be a by-product of the objective of these foreigners.

At about the time Lugard was consolidating various indigenous countries and 'English' Colonies into his Nigerian State(s), Hausaland had been under the control of the Fuduyan Fulbes for 100 years. In that time, it had regressed from Africa's richest country to one of its poorest, and has continued its slide to become the poorest country in the world today, if it were looked at on its own.

Author, George Douglas Hazzledine, penned his thoughts about the decline Hausaland had experienced around that time (1904).

> *It was a wonderful country, this Hausaland, a few centuries ago, then at the height of its importance, and it is a wonderful country now. It poured its produce and its manufactures across the desert into Europe when Europe was half civilized, and*

it pours them into the same channel still. The Morocco leather of commerce comes from Kano. Once it was as far ahead of Europe as it has now been left behind. It was once the factory of the Mediterranean, just as Morocco was the granary of Europe. Everything about the Hausa as he is to-day points to a bygone civilization long forgotten, a civilization of which his present arts and crafts are but traces.

The patterns of the workers in brass, leather, and cotton are no longer designs; they have lost their purpose, and their backbone is gone; they are but fragments of finer schemes, mere scraps of careful effects, memories of a lost art. But the brazen ewers are not savage any more than the heavier brass work of India, the hieroglyphics of Egypt, or the long gun of the Arab. They belong to a civilization, of the past, it is true, but none the less a civilization. It has left no temples, no pyramids; but then, it had no lime. Its arts and crafts have been left behind by Birmingham ; but, then, it had no machinery. It lived and survived through periods in which the Gothic invasion would have been but an ordinary incident. It was a mighty civilization, and those who built it up were, are now, and will for ages be, a mighty people.

Can we call a people savage which has a written language of its own, a language which is used to-day, and which, after much controversy, is admitted to be of older origin than Arabic ? It has been suggested that it was merely a corruption of Arabic; but that cannot be maintained, for two-thirds of it is older than Arabic. Nor can we call a people savage which has such an inborn trading instinct. The trader of Africa, the Hausa, is found now, as he has been found for a thousand years, from the Mediterranean to the Oil Rivers, from Morocco to the Nile. Everything about the Hausa indicates a past of splendour, wealth, and power.

'The Hausa has been in Hausaland longer than can be traced, and, in spite of the everlasting intertribal warfare, in spite of raiders, in spite of pestilence, has multiplied exceedingly and kept his nationality, absorbing all comers. Like the Anglo-Saxon, he gathered into his stock all the tribes and peoples, great or small, which attacked him or wandered into his land, swamping, absorbing, and assimilating them all—all except the Fulani.

English colonial policy was of course guided by its own interests and not the observations of an author, and it was in its interest and for its country's (England's) insurance for the Fuduyan Fulbe clans to continue the rule of Hausaland.

The foreign Fuduyan clans are driven by the fear of possible retribution if the bulk of the Hausa populace ever get enlightened, and thus keep scheming and actively work to prevent that happening. The control of the traditional, religious and cultural institutions (whatever of the cultural bit that has survived), has helped, as has been the ability to ability to shift the attention and minds of the Hausa populace away from its own history and rich past, and perpetuate the belief that foreigners, unbelievers, the west and (the favourite and default excuse) Southerners hate them, and are responsible for their poverty and lack. They (mostly the young men) are encouraged to go down South and kill the 'dogs' and worse. They are not leaving the job to the uninformed youth alone though, and the convergence of the three terrorist groups (Boko Haram, Fulbe Herdsmen and 'Bandits') under the 'influence' of the likes of Gumi Jr. working for the parties that control the Nigerian State[148], means that there may be up to 200,000 of these foreign militia waiting to attack indigenous Nigerian countries that have been forcefully stripped of their ability to function, defend its land and peoples, and secure lives and property.

The accounts stated by French author Bernard-Henri Levy of his visit to Nigeria in 2019, may be shocking to many around the world, but are not strange to anyone with more than a basic knowledge of Nigeria. [149]

Where Levy and most people are slightly off the mark is the assumption that there is a strict 'religious' divide among indigenous Nigerians. There is a good mix of adherents of various religions and those who are not religious at all in at least 20 (including the 'Sharia' states of Bornu, Niger, Bauchi and Gombe, and the Zuru area of Kebbi, though Kebbi is not in the count of 20) of the 36 units of the Centre (known as States) that Nigeria's dictators have created over time. This mix is sometimes found in single and extended families.

The foreign Fuduyan Fulbes must by definition continue to put up a religious front for their systematic takeover of Hausaland

[148] https://www.ghgossip.com/i-have-government-support-to-meet-with-bandits-sheikh-gumi/

[149] https://www.wsj.com/articles/the-new-war-against-africas-christians-11576880200

(even if that fails every logical test) and continued subjugation of the country.

As was the case with the 10 or more youth corpers killed in the wake of Candidate Buhari's loss, the slaughter of non-natives in many parts of fulbe controlled Hausaland over the years, often includes identifiable moslems, who are from other parts of the country and are regarded as inferior by the Fuduyan Fulbes, who consider themselves as Arabs from Saudi Arabia[150]. So when Ahmadu Bello seemingly threatens to enslave, and also come back with a sword at jeering Lagosians by saying "it is only because the Moslem power is strong here that we have not got the slaves to sell, and they are there', he is really talking about foreign Fulbe ethnicity, just like his progenitor, Mohammed Bello, who had no qualms capturing indigenous 'Nigerian' moslems in the Fulbe slave raids.

In addition to Hausaland, there were identifying and practicing moslems in many parts of Nigeria before Fuduye was hosted in the ruling courts of Gobir, with a range as diverse as the Northeastern corner of Bornu, and the Southwestern coast of Lagos.

Nigeria is now faced with the dilemma in Solomon's palace, where the foreign Fulbes would rather destroy the country than give it up to the indigenous owners, and the peace loving owners are told to give up their country for the sake of peace. [151] It doesn't matter that a couple of the numerous Fulbe Governors in Northern Nigeria has offered to give them more than enough land than they need in the heartland of Hausa country[152]. A

150 Sourced from the last pages of Ahmadu Bello's autobiography

151 https://www.vanguardngr.com/2018/07/giving-land-ranching-better-death-presidency/

152 https://www.vanguardngr.com/2018/02/gov-ganduje-asks-fulani-herdsmen-relocate-kano/

neighboring state has also previously declared that it is home for all Fulbe nomads in Nigeria and even has standing institutions to attend to all the ranching needs of the herdsmen, including the education of their children.[153]

Any continuation of the crises outside of these two states makes it clear that the objective has never been to build sustainable societies, but to subjugate and eventually (intentionally or unintentionally) destroy them.

When the Fuduyan Fulbe hold on Hausaland is eventually broken, a term more extreme than 'Stockholm Syndrome' may need to be introduced to more aptly describe the condition of 'fighting for the cause of your oppressors'.

Hausa youth mostly feel that speaking up against the obvious faults in their society is akin to blasphemy or religious sacrilege. Many Nigerians from other parts of the Nation naturally believe that questioning the structure, nature and purpose of the country amounts to treason, which is not unexpected, given that it is the only Nation that they have ever known, and information about history of the Nation is suppressed, and when available is often tainted or doctored (for instance the wild claims that the Prime Minister Tafawa Balewa was Fulbe, or that there are more migrant Fulbe in Nigeria than in the six to eight countries they are indigenous to shows a planned campaign of information tainting).

LEGIS-LOOTING

Politicians seem to have adopted a policy of "Nigeria, get yours before it is gone" and (wrongly) seem to be smartest of the lot, since it is obvious that a State conceived in, built by and sustained

[153] https://www.jigawastate.gov.ng/NomadicEducation.php

with fraud, violence and corruption will ultimately collapse on itself. You only need to listen to them speak, in privacy, about their thoughts on the country.

So as 2023 rolls by, you have politicians ready to jump into the fray to control the $15 - $40 billion the Niger Delta still provides from petroleum profits and other oil revenue, based on oil prices. A significant part of this never makes it to the treasury as the National Oil Company, and the country's Presidency gets the first bite (mostly off the records, or through ridiculous 'loss making' contracts and arrangements), so that is obviously the most desired office, with the additional powers and benefits that the approval and signing of various licenses and renewals provides.

Next in line is the Senate, which is the preferred retirement camp of Governors, and its junior associate, the House of Representatives. In most years, more is spent on the less than 500 'distinguished honourables' than the country's health and education budgets combined[154]. Former Central Bank Governor Lamido Sanusi, was in the eye of the storm as the National Assembly members feigned anger (it certainly could not be real anger) at the statement that 25% of the Federal Government's overheads was spent on/by the National Assembly.

Former House of Representative member, Gabriel Suswam, suggested that the National Assembly was ineffective and perhaps unnecessary as only about 5% of its members make any meaningful contributions at the plenary or sponsor any motions.[155] This did not stop Suswam from eventually 'retiring' to the Senate after two terms as Governor of Benue.

---

154 https://guardian.ng/issue/national-assembly-members-curious-pay/

155 https://www.vanguardngr.com/2010/12/national-assembly-overhead-when-figures-don%E2%80%99t-lie/

The official expenses are just the beginning. Nigerian federal legislators are either the highest paid or the second highest paid in the world, depending on the unstable naira exchange rate, and unraveling hidden information that should be open and transparent. Estimates range from almost $600,000[156] to $2.1 million per annum[157] for what comes to legislators for being in the National Assembly, regardless of what you want to label the payment as.

Then you have the 'constituency projects', where the legislators, supposedly due to his or her superior knowledge of the needs in his/her community, directly 'nominates' and takes delivery of items and projects for the benefit of the community. that will undertake projects to help the 'underperforming executive'. Camels and wheelbarrows are some of the odd logistics
and transport items delivered by legislators alongside the more standard motorbikes and motorized tricycles. Pepper grinders and sewing machines are also common items.

Some Constituency projects by Nigeria's Federal legislators

156 https://www.stearsng.com/article/how-much-does-lawmaking-cost-nigerians

157 https://www.change.org/p/national-assembly-of-the-federal-republic-of-nigeria-nigerian-goverment-review-and-amend-the-nigerian-senator-s-annual-pay-of-2-183-685-00-in-line-with-senator-s-pay-worldwide

According to Budgit, In 2018, over half of the projects tracked were either not done, delivered or could not be located.[158] Expectedly, when delivered and shared, it is a way to reward campaign supporters, thugs and political enforcers. In many other cases, these items lie idle in the personal homes and warehouses of the legislators for years for a wide range of reasons. Hundreds of motorcycles and tricycles[159] have been recovered in raids, as have been tractors, dialysis machines and all sorts of items and equipment.

The Federal National Assembly has however waived off any criticism of the constituency projects and the lack of proper accountability for the trillions (tens of billions of dollars, depending on the exchange rate) allocated to the program over time, and is proposing increasing the allocation to the program to 20% of the Federal Budget each year[160].

The annual budget defense of departments and agencies have also become showcases of transactional corruption. You pay to get approved, and in other cases, legislators put projects in the budget (not the constituency projects this time) to be awarded to companies owned by them or people fronting for them.

Electricity sector financial intermediary (itself an evidence of unnecessary and inefficient government intervention and corruption in Nigeria's problematic electricity sector[161]), Nigeria Bulk Electricity Trading, either mistakenly or intentionally, leaked the

158 https://yourbudgit.com/wp-content/uploads/2019/12/2018-Constituency-Project-Report.pdf (Slide 7)

159 https://guardian.ng/saturday-magazine/cover/constituency-projects-still-a-long-way-to-getting-value-for-money/

160 https://www.thisdaylive.com/index.php/2019/12/12/senators-seek-20-of-federal-budget-for-constituency-projects/

161 Generating and Distribution Companies must be allowed to take the lead and trade directly, and if necessary set up an intermediary similar to the Inter-bank settlement System, and eventually must be allowed to

fact that N7.68 billion ($15 -$19 million depending on which exchange rate used) had been added to its budget for projects outside its mandate. [162] Revelations like this have become quite routine, and mostly pass through without a hitch, except when someone feels shortchanged, or when the rare public official calls the bluff of the legislators. Expectedly, the Senate quickly hit back at the company with allegations of corruption by the management and demand that the funds be immediately returned to the Federation account[163].

Former Africa Development Bank Executive, Arunma Oteh, back in Nigeria for a stint as head of the Securities and Exchange Commission, got a taste of the wrath of the National Assembly as she accused specific members of demanding bribes and payments for trainings that were not attended. The National Assembly, ganging up to support its accused comrades, decided not to allocate any funds for the running of the Commission[164]. Ms. Oteh soon left the country to take up a position as Vice President and Treasurer of the DC-based World Bank Group.

---

engage in corporate action that will enable organic vertical mergers and acquisitions in the sector.

162 https://www.thisdaylive.com/index.php/2021/07/14/we-didnt-insert-n7-68bn-in-our-budget-says-nbet/

163 https://dailypost.ng/2021/08/08/senate-tells-nbet-to-refund-n132-million-spent-on-drivers-uniforms/

164 https://thenationonlineng.net/zero-allocation-for-sec/

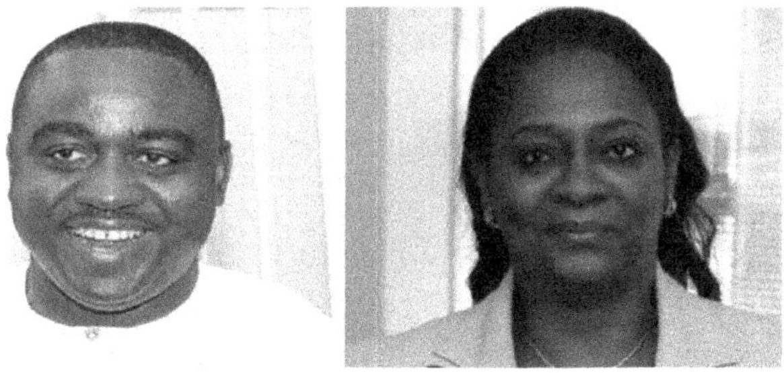

L: Suswam-Described most legislators as idle and ineffective R: Oteh-Clashed with legislators as Head of the SEC

We'll not bother with small matters like Senators forcing private companies to sponsor them to events like the 2010 World Cup, since some of these events (including the general organisation by and conduct of Nigerian officials) are too painful and shameful to document.

The obvious fact is that a nation with Nigeria's history and Constitution (Constituent Countries and not the much criticized document) does not need a 'National' Assembly, and should at best be a very loose confederation between willing constituent units. The strange reality is that those shouting 'one Nigeria' by all means and at all costs are mostly a strange collection of terrorists, foreign (and non-indigenous) violent propagators, extremely corrupt politicians, the very ignorant, and those who cannot by any reasonable measure can be referred to, as men or women of goodwill, is very good evidence that the country as currently structured, is deeply flawed, fraudulent and doomed for catastrophic failure.

If there was ever any dou,bt about this, you only need to look at the outcomes. The once richest country in Africa has become the poorest part of the world under foreign rule; and if you think the future is going to be any better, you only need to look at the Children. The leaders (apologies for using that word to describe office occupants and title holders who corner funds they did not earn and have no idea how to build sustainable societies) shamelessly argue whether there are 7 or 10 million children out of school[165].

Those are millions of children that may not lead productive or useful lives under the current structure where foreigners, whose only objective is exploitation and subjugation of the country by whatever means (religious or not) oversee a system of subjugation of indegious peoples leading to a lack of accountability and a breakdown of the family structure (Al-Majiri etc) and the destabilization of previously sustainable societies.

Unless there is a drastic change, the terror tripartite of the 100,000 terrorists of Gumi Jr., alongside their Boko Haram Partners and the Foreign Fulbe Herdsmen are a minuscule fraction of what is looming.

This needed and drastic change is not in the interest of the foreign ruling class though, due to the belief that an uneducated and unproductive populace is easy to manipulate and control. You thus have a 'rule by force' mentality as the foreigners do not mind burning the country to the ground, since they can easily pack up and return to their own countries, as the various clan leaders were doing, as Lugard's boys closed in on various cities in Hausa country, until they were invited back. Rather than attempt to return the 'conquered lands' to sustainable country, you have situations where these foreigners keep inviting terrorists into the country to consolidate their hold166, with an aggressive identification scheme quickly 'Nigerianizing' people with no cultural, ethnic or historical links to any indigenous Nigerian ethnic

165 https://guardian.ng/opinion/out-of-school-children-still-depressing/

166 https://www.thisdaylive.com/index.php/2019/09/17/bauchi-gov-nltp-will-accommodate-fulani-herdsmen-from-neighbouring-countries/

group, while threatening the rights and ability of indigenes of constituent Nigerian countries who refuse to be subjugated by the fraudulent identity scheme of foreigners to function, exist and live normal lives in their own countries.

GOVERNORS

Sitting right below the unneeded 'National Assembly' in terms of access to the mostly Niger-Delta generated treasury are the overlords of the 36 geographical branches of Nigeria's central or 'Federal Government', the Governors of the States.

Rather than natural or autonomous component units of willing and voluntary union, the Nigerian 'States' are mostly arbitrary and random creations by young, inexperienced and sometimes crazed dictators, with some of the 'stated' (pun unintended) reasons including to reward sycophants or gain the tactical advantage and take the initiative from perceived or real enemies.

The Nigerian Senate is reported to be considering the creation of 20 new states[167] to put an exclamation mark on the transient and ridiculous nature of the show[168]. This is radically different from the expansion of the United States, with new territory coming into the union.

The 'creation of States' is part of a concerted and determined attempt to suppress and decimate the constituent countries that are wholly or partly subsumed in Nigeria, which are very different in culture and historical development. The creation of 'homogeneous' states is nothing short of lunacy, except that it is

[167] https://www.bbc.com/pidgin/tori-58120932

[168] https://punchng.com/full-list-senate-gives-condition-for-creation-of-20-new-states/

an attempt to further weaken the ability of the natural countries to function. These functions should naturally include securing lives and property, promoting welfare and the conditions for individuals and businesses to thrive, ensuring justice and the law, providing public goods and services, and ensuring sustainable societies and not one of dependent beggars and corrupt thieves.

The current situation, however, works for the interests of foreigners, terrorists, extremists and invaders who have their own countries apart from the constituent countries of Nigeria, but in whose interest the Nigerian State (in collaboration with another set of colonials) was set up, for the exploitation of the indigenous countries. The deliberations on the very easy solutions that will change the seemingly doomed course of the people of Nigeria cannot include this group, but should also not include group retribution, punishment or any form of crimes or illegality.

It is therefore not surprising that States that were created to fail are not sustainable, and almost completely dependent on stipends from the Centre (i.e. Niger-Delta oil money) and can only manage to pay salaries (most times) with a significant portion of the allocation diverted or embezzled by the Emperor-Governors, their Godfathers and Associates.

As previously mentioned, discussing individual acts of corruption is almost redundant, and quite distracting from the real issues, in the context of the Nigerian State as currently structured. It is probably worth mentioning that apart from seeking safety in the Senate, former Governors can easily have $80 million embezzlement charges disappear by switching to the right (or ruling) party, aided by the 'the Constitutional right of the Attorney General' to withdraw the criminal case.[169] It also doesn't hurt that the Attorney General's cousin heads the office

169 https://punchng.com/why-i-withdrew-gojes-n25bn-fraud-case-agf/

of the role duplicating waste of time and money, that is packaged as an anti-corruption agency that is mostly very effective at hounding political opponents, and then tries to gain credibility by raiding night clubs and arresting young men en masse.

Little wonder that there is little left in terms of funding for infrastructure, and successful entrepreneurs and independent voices are sometimes seen as threats who need to be shut down, especially if they have any opinion on governance, or perceived interest in politics. In mid-2021, the Federal Government banned twitter in the country and Pantami's former PA, assuming regulatory powers over the technology sector, seemed to be scheming a crackdown on technology and e-commerce companies that have developed in spite of the corruption and apathy in Nigeria[170].

Consequently, many bright young people, who should provide the brainpower needed to develop the Nation, feel that they need to escape abroad, even if for the most menial of jobs, to escape the madness and nightmare that is playing out in front of them.

In addition to seeking cover in the Senate, some Governors still try to hold on to the treasury and control that comes with the office. The default move was to install a stooge (with crazy oaths and shrine ceremonies involved in ensuring continued allegiance), but this has had limited 'success', with one or two notable exceptions.

A reported recent trend has been for the Governor to sell the ruling party ticket and, through his control of the state party machinery, ensure the transaction leads to the emergence of the party's candidate. The going rate in recent electoral cycles

---

170 https://techcrunch.com/2021/08/17/a-leaked-bill-for-nigerian-startups-reveals-a-theme-of-licenses-fees-fines-and-sentences/

has ranged between $5 million and $10 million, and it is not surprising that executives of the top public fiscal, monetary and National Oil companies have been listed by some sources as recent buyers.

While top federal public officials remain the primary beneficiaries of the billions stolen through the forex regime, petroleum subsidies and the loose fiscal regime, executives at these public institutions are more than happy to be incidental beneficiaries and sometimes enablers of what might seem to them as a victimless crime, but is essentially the continuous pillaging, looting and destruction of indigenous Nigerian countries and peoples.

Media and Construction mogul (and Buhari regime Insider) Isa Funtua discusses matters of interest with Central Bank Governor, Godwin Emefiele, with Abdulrahman Danbazzau close by.

As such, the foundations, intent and philosophy of the Nigerian States and the structure imposed by their existence is meant to impoverish indigenous Nigerians and weaken their identity, and as such the outcomes cannot be different from their intent.

Just like the executives of public enterprises mentioned above, civil servants (jokingly called 'evil servants' by those that routinely have to deal with them), are sometimes beneficiaries and many times enablers of the decay and corruption of the Nigerian State.

Philip Asiodu's comments and hypotheses on the start of the corruption in Nigeria civil service may or may not hold water, but there is universal agreement that there is widespread corruption in the civil service and the rest of the public service. This is why Ministers and political leadership can be an opportunity (not a certain way) to moderate the ineffectiveness and corruption, and also why Buhari's comment about not needing Ministers revealed a bankruptcy that had little to do with his claims of having no money to buy campaign forms.

Many insiders know that many times you see a civil servant working hard on a 'file' or project, it is likely because he has been financially induced or has a financial involvement in whatever is being pushed. It is of course unfair to paint everyone with one brush, since honest hardworking civil servants exist, swimming against the tide, due to personal conviction, or the rare case of departmental or institutional leadership. It must however be emphasized that in many aspects of Nigerian public life, choosing not to participate in the seemingly established way of 'doing things and taking care of everybody', can put the individual at risk, especially in institutions like the police. Also, to a large extent the civil service problem is one of the national structure and not individual character.

A minor point regarding the current civil service in Nigeria is that the attitude of most of the rest of the country is quite different from that of the Fulbe ruling clans of Hausaland. Almost 50 years after Asiodu's referral point, civil servants from other parts of the country are often viewed as people who could not go abroad or get jobs in leading private institutions. The reverse is somewhat the case in Fulbe controlled Hausaland, where obviously, the country's religious, cultural and political institutions are controlled by a few foreign clans. The civil service (and the public sector in general) is extremely important for the control of the Nigerian state and the exertion of influence. It is therefore, quite important to the clans to ensure that the leadership and other key positions in the civil service and public institutions have critical numbers from the ruling clans, and of their associates. As such, while people from other parts of the country are mostly indifferent about appointments of Permanent Secretaries, senior civil servants and public positions, these appointments are followed with dogmatic interest and reactions by members of the Fulbe ruling clans.

LABOUR

The labour unions do not have a constitutional right to allocations or disbursements from the federal treasury, but they compete for the resources and exert tremendous influence as gatekeepers, and sometimes as agents of destabilization in national affairs.

Unions, by definition, should and will seek to protect the interests of their members (ideally in a legitimate and ethical way), but, perhaps, due to the structural issues that face Nigeria and other factors, the labour unions compete, alongside politicians and civil servants, for the resources of the country in an aggressive and oftentimes damaging way.

The Unions have been used by the National Oil Company and other organizations in preventing much needed reforms that would have struck at the heart of corruption and improved the employment numbers and situation in the country.

Labour Unions have also been used as weapons in political battles, and a Governor accused his colleague from a neighboring state of paying bribes to labour leaders and sponsoring demonstrations in his state[171].

Several examples exist in the private sector, where labour leaders, like the politicians and legislators, force private organizations, especially those in their industries, to part with gifts, cash and other benefits or sponsorship, or face the risk of the unions descending on them with picketing and other business and operations disrupting activities.

The unions are known to violate the basic requirements they recommend in dealings with their own employees[172], and oftentimes do not strive for economic improvement or the creation of the atmosphere for growth and development, which ultimately lead to greater availability of jobs and improved welfare for workers.

Factors of production have to be priced fairly, for production to be possible and sustained over the long term. The labour unions, who are collaborators with the National Oil Company in massive corruption, would rather have billions of dollars wasted and stolen in 'maintaining' the national liabilities known as refineries, than have the massive corruption ended.

If only it were possible for these labour leaders and unions to be given loans and allowed to manage one refinery or critical assets they insist on being run unprofitably and inefficiently (or actually not run at all), and their compensation was completely dependent on the earnings, so they can understand that sustainability and good practices are essential for economic growth

---

171 https://gazettengr.com/audio-el-rufai-accuses-ganduje-of-bribing-nlc-leaders-to-cause-mayhem-in-kaduna/

172 https://www.thccable.ng/nlc-yet-to-implement-minimum-wage-for-own-staff-2-years-after

and, hopefully they will get to see first-hand that wasting billions maintaining unproductive refineries and other 'public' assets are tantamount to massive corruption with identifiable pecuniary beneficiaries, and against everything that an incorrupt and remotely knowledgeable labour union should stand for.

This, in essence, is the history of the Nation the Nigerian military helped create and has, sometimes forcefully held together and, at other times, tried to break up violently. It is not

on its current trajectory.

Most of the time, it has just been a mechanism by foreigners to loot the resources and subjugate the indigenous constituent countries. Its rich countries and amazing indigenous peoples, with some of the richest and most diverse cultures, are currently trapped in a Nation, set up with no consideration for them, and without their consultation.

Indigenous Nigerians, especially the young, must legally and methodically flip the script. However, they should not do this by participating in violence or the systematic destruction of their own countries. That would be playing into the hands of those whose objectives have always included the subjugation or destruction of the countries, in the first place. This is why the recent strategy (2021) by IPOB and the ESN is problematic and could be counter-productive.

Yet, the indigenous Nigerians must realize that the good of their indigenous peoples and countries MUST come first, not that of the Nigerian state. Certainly, not that of the foreign Fulbe ruling clans who can quickly beat a retreat to their indigenous countries where they still exert significant influence; and not that of the old colonials and complicit Europeans who may be trying to cover up their (predecessor's) past crimes and are worried about possible demands for reparations (this is primarily ineffective and should be discouraged as much as anyway). And finally, not that of an international order that has shown little concern or consideration for the fate of the Black African.

The pursuit of the interests of the peoples and indigenous countries of Nigeria must be prioritized, and this must be done legally and peacefully.

REFERENCES

Adegbamigbe, A. (2020, March 5). Retrieved May 19, 2021, from The News Nigeria: https://www.thenewsnigeria.com.ng/2020/03/05/34th-memorial-how-ibb-executed-vatsa-wife-narrates-last-moments/

Akinbode, A. (2020, April 25). *National History: Why the 1992 Gideon Orkar Coup failed*. Retrieved May 19, 2021, from History Ville: https://www.thehistoryville.com/gideon-orkar-coup/

Daily Trust. (2019, March 3). *Reminiscences with Gen Ipoola Alani Akinrinade*. Retrieved from Daily Trust: https://dailytrust.com/reminiscences-with-gen-ipoola-alani-akinrinade

Diamond, L. (1988). The Crises in Western Nigeria. In *Crisis and Conflict in the Western Region, 1962–63. In: Class, Ethnicity and Democracy in Nigeria*. (pp. 93-94). London: Palgrave Macmillan. doi:https://doi.org/10.1007/978-1-349-08080-9_4

Ekine, S. (2010, May 10). *The mysterious disappearance and death of Nigeria's President*. Retrieved May 19, 2021, from The Globe and Mail: https://www.theglobeandmail.com/

news/world/the-mysterious-disappearance-and-death-of-nigerias-president/article4318360/

Ekine, S. (2010, May 10). *The mysterious disappearance and death of Nigeria's President.* Retrieved May 19, 2021, from The Globe and Mail: https://www.theglobeandmail.com/news/world/the-mysterious-disappearance-and-death-of-nigerias-president/article4318360/

Obiejesi, K. (2018, July 18). *iNews: Mandela begged Abacha not to execute Ken Saro-Wiwa and companions.* Retrieved May 19, 2020, from International Centre for investigative Reporting: https://www.icirnigeria.org/mandela-begged-abacha-not-to-execute-ken-saro-wiwa-and-companions/

Oduyela, S. (2004, September 17). *Owners of Nigeria - Part 4.* Retrieved from Dawodu.com: https://www.dawodu.com/oduyela9.htm

Oduyela, S. (2004, September 17). *Owners of Nigeria Part 4.* Retrieved May 19, 2021, from Dawodu.com: https://www.dawodu.com/oduyela9.htm

Oduyela, S. (2004, October 21). *Owners of Nigeria Part 8.* Retrieved May 19, 2021, from Dawudu.com: https://www.dawodu.com/oduyela13.htm

Oduyela, S. (2004, October October). *Owners of Nigeria: Part 6.* Retrieved May 19, 2019, from Dawodu.com: https://www.dawodu.com/oduyela11.htm

Omoigui, N. (2021, May 19). *History of Civil-Military Relations in Nigeria (5): the Second Transition (1979-83, Part 2).* Retrieved from Dawodu.com: https://www.dawodu.com/omoigui7.htm

Omoigui, N. (n.d.). *Col. Dimka's Failed Coup Attempt.* Retrieved May 19, 2021, from Urhobo Waado: http://www.waado.org/nigerdelta/nigeria_facts/militaryrule/omoigui/Dimka-1976.html

Omoigui, N. (n.d.). *Military Rebellion of July 29*. Retrieved May 19, 2021, from Dawodu.com: https://www.dawodu.com/omoigui45.htm

Omoigui, N. (n.d.). *Murtala Ramat Muhammed (1938-1976)*. Retrieved May 19, 2021, from Dawodu.com: https://dawodu.com/murtala3.htm

Oredein, O. (1985, April 10). *Three Nigerians shot for drug possession*. Retrieved from United Press International: https://www.upi.com/Archives/1985/04/10/Three-Nigerians-shot-for-drug-possession/7864481957200/

Siollun, M. (n.d.). *The Rollercoaster Life of Murtala Muhammed*. Retrieved May 19, 2021, from Dawodu.com: https://www.dawodu.com/siollun2.htm

The New York Times. (1976, March 13). *Nigeria Executes 30 for Coup Role*. Retrieved from The New York Times: https://www.nytimes.com/1976/03/13/archives/nigeria-executes-30-for-coup-role-exdefense-minister-among-them.html

Walker, A. (2016, February 4). *World News*. Retrieved May 19, 2021, from The Guardian: https://www.theguardian.com/world/2016/feb/04/join-us-or-die-birth-of-boko-haram

Watch, H. R. (2002, April 1). *Nigeria: Soldiers Massacre Civilians in Revenge Attack in Benue State*. Retrieved May 19, 2021, from Human Rights Watch: https://www.hrw.org/news/2001/10/25/nigeria-soldiers-massacre-civilians-revenge-attack-benue-state

PART 4

The Weak Ties that Bind, Have become even Weaker

This section takes a brief look at how the weak ties that bind the Nigerian State together have become even weaker, raising the obvious questions regarding whether

1) Nigeria is a failed or failing State
2) The lives and condition of the indigenous peoples and constituent countries would be significantly better in a post-Nigeria or Confederation of Nigerian States arrangement with different constitutions for the constituent Countries.

Each country in the United Kingdom, for instance, has significant control in various matters and even have different national assemblies and sports teams. They choose to present different teams in the World Cup, but one team at the Olympics. On the other hand, Puerto Rico, a territory of the United States, presents its own separate team at the Olympics.

These things can easily be worked out.

The Military, Crude Oil, ignorance and suppressed information, are some of the critical factors that have held the Nigerian State together, especially since 1970. The influence and interest of the British Empire was the dominant factor in the prior decades, but has naturally waned as the might of the empire declined. From the fantastically corrupt description by the fantastically corrupt[173] David Cameron to his esteemed audience, it appears that the British Elite doesn't remember that Nigeria is (almost) fully its creation. The Head of the English Church also has a strange urge to defend the indefensible regarding Nigeria, perhaps due to a nostalgic fondness of his time in the Oil industry.

173 https://www.ft.com/content/4362e62f-00fc-4cd9-a20b-9134bc2f0699

CHANGING COLONIAL INTERESTS AND PRIORITIES

In some cases, England and France found it efficient and justifiable to use a human ranking system for the execution of their plunder from Berlin (Colonialism if you prefer), and even in the few decades prior. The African was too savage to deal with directly or even convert to their 'European religion', so clans identifying as Arab or semi-Arab were to be used (sometimes after light military pacification) to do some of the dirty work, and in some cases, 'partially civilize and proselytize' the 'savage' Africans (even if this involved mass murders and slave raiding).

There are a few problems with this position, but the one to be highlighted here is the fact that right around this time, Western European philosophy had largely reached a consensus and conclusion against the existence of God, and the need to prevent a labor crisis from a potential massive unemployment of clergy shifted theology's focus to humanism, and the dignity of man, at least of the European Man.

As such, the Colonials, for the most part, could not give what they didn't have in terms of God, and focused on exploiting the colonies and its resources. In Nigeria, British Colonial focus changed quickly to match its interests, moving from countering the French (in Borgu for instance), to ensuring the status quo was maintained, so Nigeria (Hausa Country under Fulbe rule) could be the main source of troops if WW3 broke out. Crude Oil became a big factor around the time of the Nigeria-Biafra civil war occurred, but the emergence of an Africa-first foreign policy (backed by petro-dollars) meant that Nigeria under Obasanjo could 'nationalize' British Petroleum's holdings in Nigeria, for supplying Crude Oil from the Niger-Delta to the Aparthied South African Government in violation of International sanctions.

From the 80's onwards, England mostly had no special interests in Nigeria, as it focussed on European integration, but with Brexit, England may have Nigeria in its sights again, although it is unlikely that its faltering and unstable creation can be of much use to it.

The 'rulers' of that creation, the Fuduyan Fulbes however view the English crown as the source of their authority over the indigenous countries of Nigeria, and adore Lugard, often meeting up to discuss their interests and plot their strategies in the unsurprisingly named Lugard House in Kaduna.

In a weird twist of fate, Matthew Dennison, the biographer of the universally adored English Monarch, Queen Elizabeth, while reacting to allegations of racism within the royal family's own ranks, chose to turn to Awolowo in the paragraph concluding his anti-racism defense. The same Awolowo that suffered greatly for failing to do the bidding of the British in the 50's (and ended up in jail, with his comrades in the 60's). The same Awolowo, that pretty much broke the back of English control by forcing BP and the oil companies to pay royalties to the Central Government[174].

Things sure do have a way of coming full circle.

England's frenemy and neighbor, France has done more than its own fair share of damage in Africa, particularly in West Africa, where it still has major economic control with the CFA franc, and whose role in the security situation in the Sahel, especially with Arab identifying groups that claim a disturbing right and mandate to slaughter indigenous Africans in their own towns and villages, remains quite suspect.

Two events playing out in 2021 in Guinea and Mali should be observed.

Alpha Conde was overthrown in Guinea, by his French trained Chief Bodyguard, which has naturally ticked Russia, and its mining interests off. Any transition announced is expected to result in the long awaited election of a Fulbe head in their main homeland, which may be linked to some sort of deal between the Nigerian based Fuduyan Fulbes and the French.

This may have affected Russia's commercial arrangements with Conde, and Russia commercially[175] offered neighboring Mali security help, which drew the ire of France, considering all the military 'help' it was already offering to Mali to fight the Arab identifying groups that claim a right to kill Africans at will. France has expectedly been throwing tantrums[176] at this development, but it is obvious that reducing the influence of France in the region, as well as its allied Arab identifying extremists,

---

174 This wartime decision was a disaster for Federalism, and should have been reversed once the wartime conditions ceased

175 https://www.aljazeera.com/news/2021/9/25/mali-approached-russian-private-companies-moscow-not-involved

176 https://www.rfi.fr/en/france/20210916-france-angry-prospect-wagner-russia-mercenaries-training-soldiers-mali-junta

will be for the good of the core West African homeland of black Africans.

Russia, like the United States, has, thus far, had little colonial or exploitative interest in West Africa, so their actions may be less suspicious than the traditional colonial powers. However, indigenous African leaders must start defining the interests of their homeland, and start working for those interests.

The United States has mostly trusted English or French foreign policy for its limited interactions and interventions in West Africa, but has lately needed to go on its own to avert disaster, or protect its citizens. Despite a firm silence on the matter, it is believed to have resolved the Abacha-Abiola situation, which led to the emergence of Obasanjo as President, and helped temporarily stabilize a floundering Nigeria.

More recently, the rescue of an American Citizen Philip Walton, likely from the hands of Fulbe bandits in Nigerian territory[177] provided a jolt for the Nigerian government, and a Military, that has admitted to once waiting for approval to engage foreign Fulbe terrorists who were killing indigenous Nigerian citizens.

MILITARY

It was a largely insignificant "blink and you miss it" moment that played out as Nigerian Army Chief, Lt. Gen Farouk Yahaya visited troops preparing to turn up the heat (some say due to US pressure) on the terrorists that had been running free on Nigeria's Northwestern zone, and seemed set to make southward incursions to Nupe Country.

---

177 https://www.reuters.com/article/us-usa-nigeria-military-idUSKBN27G0JI

As observed on the Channels TV News, General Yahaya was with a unit that seemed to have between 20 and 50 men, and engaged in light banter in Hausa, but the lack of response from the group made him question if anyone in the group spoke Hausa. Still no response.

Even ignoring or considering suggestions of profiling of men and officers sent to the frontlines, it was hard to imagine that it was tough to find a Hausa speaker in this mixed police and military group in Zamfara, the heart of Hausa Country. The Nigerian military was founded and has largely remained a 'Hausa' organization, with most of the training (Defense Academy, Zaria Depot) concentrated in Hausa Country. Other single campus institutions like the Nigerian Law School were turned to multi-campus schools following agitation for the same. It is obvious that the military training institutions, including the flying training schools (in Kaduna and Kano) will have to follow the same trend in a Confederation of Nigerian States.

Lt. Gen Yahaya himself became Army Chief after dozens of his superiors were retired to clear the path for him as a candidate acceptable to the President, following the tragic death of his predecessor, Lt. Gen Attahiru.

It is understandable if there is no abundance of Fulbes in the Nigerian Military, given that Fulbes likely make up less than 1% of Nigeria's population, but are Governors of at least 8 of the 10 core Hausa States and have a preference for Nigeria's 'revenue generating' (or in some cases revenue diverting) agencies.

However, not having Hausa speaking soldiers in a unit of at least 20 is shocking. It might be an aberration, but it is quite telling.

That the Nigerian Military has seemingly struggled to put down insurgencies raging the country is not due to capacity. During the Nigeria-Biafra War, the political (or fiscal) leadership ensured divisional commanders were responsible for the purchase of weapons. Lately, Nigeriens (see blame it on Niger), and ranking Fulbe clan leaders from within the Military's own (retired) ranks, have been revealed to be involved in weapons funding scandals.

Quelling insurgencies go beyond weapons though, and the Nigerian Military must acknowledge its role in forcing and forging indigenous countries into the very unnatural and inefficient colonial arrangement known as Nigeria.

The Nigerian Army itself still bears the insignia of the invading Fuduyan Fulbe foreign militia.

Special consideration must be made for the Military (funding etc), as indigenous countries work out the best arrangement for a Confederation of Nigerian States, but the Military must also now stand by and play a watchful, but non-interference role as these arrangements are worked out.

The Military's involvement in the formation of Nigeria was to the benefit of foreign colonials. It must now work for the indigenous peoples and countries of the Nigerian State.

CRUDE OIL

Crude Oil was first drilled in commercial quantity in Nigeria in (Oloibiri) in the Niger Delta region of Southern Nigeria in 1956 by Shell-BP. By the 60's, Crude Oil was gaining relevance over Palm Oil as Nigeria's most important oil and commodity.

Though Crude was not a causative factor of the Biafra-Nigerian Civil war, it proved to be a major consideration in the International politics and alignments that had some Western powers either directly or covertly supporting either side in the war.

Nigeria quickly joined the Organization of Petroleum Exporting Countries (OPEC) shortly after the end of the civil war and was able to benefit from the OPEC oil embargo, which quadrupled oil prices. The impact of the increased prices were multiplied with an increase in production and Crude oil, which accounted for less than 1% of all Nigerian exports in 1971 made up a stunning 97% in 1984.

While production was limited by its membership of OPEC, Nigeria achieved a peak production average of 2.5 million bpd in 2004, a long road from the 5,100 bpd extracted from its first commercial well in the late 50's.

The focus on, and increase in crude oil revenues, led to a neglect and frankly, abuse of both the other productive sectors of the economy, and strangely, the communities in which crude oil wealth was generated.

Crude Oil also helped Nigeria become a strong, central, top-down state that effectively seized power from the constituent nations and ethnicities that lie wholly or partly within Nigeria. The country effectively became a unitary state, whose leadership was often dependent on the success of military coups, and whose dictators created states by decrees and fiats to maintain some false semblance of a federation. As such most Nigerian states, apart from being economically unviable, were set up without adequate consideration for the historical and cultural facts and situations.

These states are effectively a department of the Nigerian corporation or experiment, rather than independent, natural building blocks and constituent units of a federation.

The country then created a National Oil company, the Nigerian National Petroleum Company (NNPC) in 1977. NNPC has replicated the experience of the Nigerian government by largely existing for its own benefit, and its impact is generally thought to be value-negating for the country.

Almost all of Nigeria's crude is extracted from the Niger-Delta region, and in terms of the current configuration of Nigerian states, Nigeria has four major oil producing states (Akwa Ibom-31.4%, Delta – 21.56%, Rivers – 21.43%, Bayelsa – 18.07%, and five minor oil producing states Ondo – 3.74% Lagos – 2.64%, Edo – 2.06%, Imo-1.06 % and Abia-0.68%.

In the 50 year post civil war period, especially since the OPEC oil shocks, crude revenue has been the financial glue by which the emergent unitary state held its sub-divisions together. Oil revenue has contributed over 50% (over 60% in most years) of government revenue in the last few decades.. These revenues are 'shared' in a monthly FAAC (Federation Account Allocation Committee) in which the Head Office of the Nigerian nation disburses funds to its field units (states).

Crude oil is also said to provide about 95% of the Government's foreign exchange earnings, but the fact that Diaspora remittances have recently matched or outpaced crude oil earnings, could somewhat de-emphasize the significance of this point, if the Central Bank removes barriers and allows transparency and the efficient use of technology.

Crude Oil and the Petroleum industry also account for between 5% and 10% of GDP in recent years, although the volume of activity would have been much higher, but for OPEC membership (upstream) and the existence of subsidies and other distortions and forms of corruption, which ensures an import driven supply for the market in the downstream sector.

Year	Daily Average Oil Production in Nigeria *(in 1,000,000 barrels)*
2019	2.109
2018	2.007
2017	1.969
2016	1.900
2015	2.201
2014	2.276
2013	2.279
2012	2.412
2011	2.461
2010	2.533
2009	2.211
2008	2.172
2007	2.208
2006	2.433
2005	2.527
2004	2.430
2003	2.233
2002	2.087
2001	2.268
2000	2.159
1999	2.066
1998	2.167

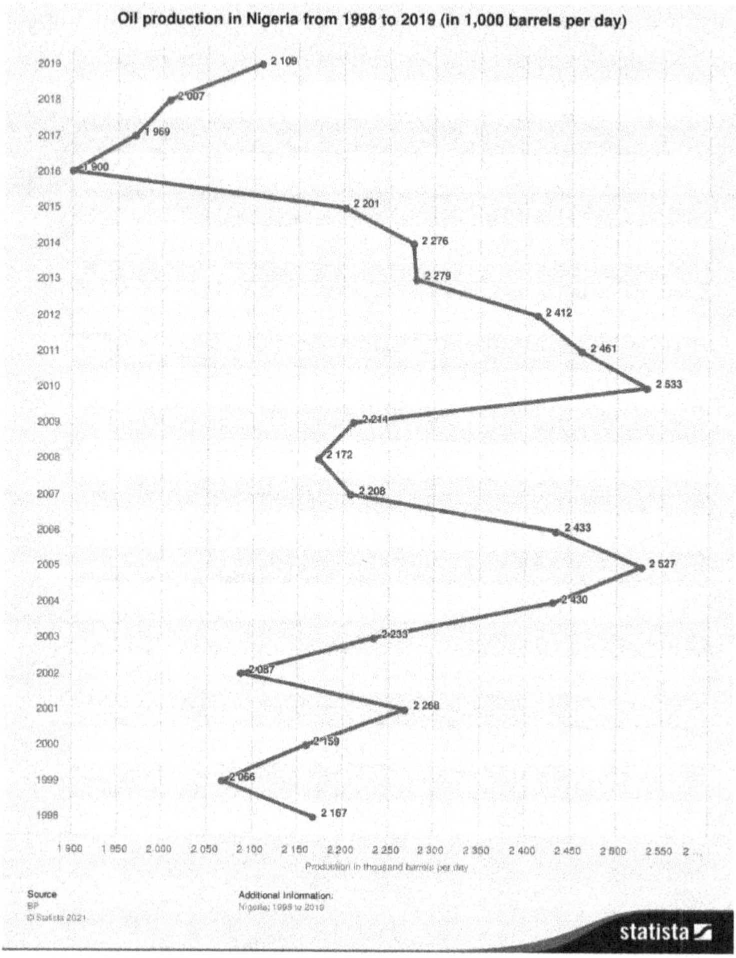

Oil production in Nigeria from 1998 to 2019(in 1,000 barrels per day)[178]

After the Biafra war, there was an increase in the global oil price, but there certainly hasn't been an oil boom for most of the indigenes of Nigeria with the corruption, inefficiency, and

[178] (Sönnichsen, Nigeria's oil production in barrels daily 1998-2019, 2021)

limited transparency of the sector, especially since the emergence of the National Oil Company (NNPC), which on paper helped strengthen the control of the sector by Nigeria's central government, but functionally has been a disaster for transparency, and a fertile environment for corruption.

The guaranteed stream of revenues over the years have however been a means for Nigeria's (legal or functional) ruling class to maintain its hold on the country.

Crude oil has been a crucial binding string of the Nigerian State, since Awolowo forced BP to pay royalties and taxes solely to the Central Government of Nigeria during the civil war.

It can be argued that this was a wartime necessity that should not have become the status quo.

The window on the crude oil boom is closing with demand (electric cars, renewable energy etc) and supply (new extractive technologies, sanctioned sources coming to market) pressure suggesting that Crude Oil may have less than a decade to be able to command abnormally high prices.

It may be time for the producing communities to organise 'community trusts' to stake a direct claim for royalties, especially with NNPC's 'reorganisation' as a private company. The (Niger-Delta) states are not best suited for this, but can aid ethnic nationality groups like the Ijaw National Congress, and their Ibibio, Annang and Urhobo equivalents to take the lead on this.

Other peoples in the Nigerian state have taken control of their extractive minerals like gold and lead. The communities in the Niger-Delta, who have suffered so much environmental damage, should not be made to wait for a band-aid like the Niger Delta Development Commission. Community-owned endowments can then be formed to tackle the environmental damage and work towards sustainable development of the oil producing ethnic nationalities.

NIGERIA'S FOREIGN FULBE QUESTION: THE OLD, NEW, PART AND WANNABE COLONIALS

The bolded quote below, is from a facebook post from defender of all things Fulbe, Dr. Aliyu Tilde. While he alludes to the fact that the Fulbe, whose home Country is at least 1000 KM away from Hausa Country, control the traditional, religious and political institutions in Hausaland, it is interesting that he claims that outsiders are trying to 'separate the Hausa from the Fulani' when most outsiders do not know or care about the difference and see them as people from the same place speaking the same language.

Most outsiders get the first inkling of a difference due to the sheer disgust displayed by the Fulbe ruling clans, when innocently and ignorantly described as Hausa. It is almost surreal to see such level of resentment shown towards a group whose Country has been subjugated, but it is probably not that uncommon, given the 'Apes Obey' example (A short history of the Nigerian Military)

'Under the Fulani, Hausaland became united and expanded both its geography and influence right up to the Cameroons. It is that unity that saved it from a second attempt of subjugation by the Kanuri in the early 19th Century. It is that unity that played a vital role in modern Nigeria, saving the North from domination of some southerners who planned to subjugate the region when the Europeans left. That vision of subjugation was frustrated and will remain so forever.
By coincidence, the defence took place under the leadership of the Sardauna, a Fulani prince from Sokoto. The Nzeogwus and FFKs[179] *did not miss that.*

The attempt to separate the Fulani from the Hausa is, therefore, a strategy to weaken that unity. We must not aid it.

Before I leave, I must confess that Fulani leaders today are not taking this role of defending Hausaland seriously. Most of them have indulged in material decadence in consonance with the Khaldunian thesis of cyclic pathway of civilization. Especially when learned, a Fulani must always strive to live by the code of pulaaku: ascetic, altruistic, generous, shy, independent and courageous. That life of sacrifice for the common good is what gave them victory yesterday and made it possible for them to survive in many climes.

The lifestyle of many of us today violate the pulaaku code in many fundamental ways. We loot public treasury and pursue meaningless luxury at the expense of the majority, as do others. In spite of this gap, the Fulani can still be counted upon by their neighbours when the need arises

[179] FFK here refers to Femi, the son of Remi Fani-Kayode, who combined with Ladoke Akintola to cause the confusion in the Action Group in the early 60's. Femi has had at least two separate spells in both of Nigeria's leading political parties.

as recent events tend to show. Man can be distracted by the material but his DNA remains the same.

So let the 1966 coupists, Biafrans and FFKs continue with their castigations. That will not change the Fulani genetics one bit. As for the Fulani relationship with Hausas, tib-da-taya, mutu-ka-raba.'

Dr. Aliyu U. Tilde
24 July 2020

Dr.Tilde in this writeup suggests that the Fulbe are liberators of Hausa Country from Kanuri and 'Southern' invaders. Perhaps he has not read about Modibbo Adama's quest to go and kill and plunder peoples and countries that had done him no wrong, or Mohammed Bello's killings and slave raidings that destroyed the lives of millions. Under Fulbe rule, Hausaland quickly reversed its status from Africa's richest country to one of its poorest, and has remained so for over a Century.

Dr. Tilde's description of the unity between the Fuduyan Fulbes (most Fulbes are in their home country in the Nations of Guinea, Guinea-Bissau, Mali, Senegal, Gambia and Mauritania, so the distinction is necessary) and Hausa has the Fulbes as Governors in at least 8 of the 10 or so states in the Nigerian section of Hausa Country, despite making up anywhere between 1% and the almost impossible; but claimed 6% of the population. Great compensation for the defense of Hausaland, but it sounds like the type of situation God warned against, suggesting that a people should not let foreigners rule over them.

Not that it matters to the rest of the countries in the Nigerian State, who are desperately trying to get away from Fulbe-controlled Hausa Country and their sometimes (especially recently with Buhari) willing collaborators from Kanuri

Country. Herdsmen, terrorists and drifters, almost all foreign Fulbe are wrecking havoc all over Nigeria, and the populace is unsure whether this is a grand design at work, or just a 'beast' that its creators have lost control of.

The claims of superior DNA and race will be ignored, as groups of people are free to say or claim whatever makes them feel good about themselves, though it should be pointed out that compared to the rich culture and values intrinsic to many indigenous parts of Nigeria, it may not be farfetched to respectfully refer to Fulbe culture as still in the process of being refined.

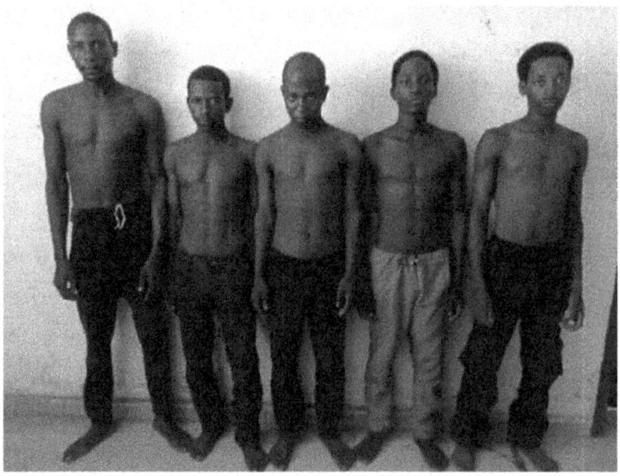

Superior DNA leading to 70% (in this case 80%) of the kidnappings, in places far away from the Fulbe defense of Hausaland.[180]

Beyond claims of superior DNA, ethos and culture to more serious matters, Tilde admits that some Fulbe leaders loot the public treasury, but he may be understating the problem, given the influence, overrepresentation and some would say dominance of the ruling Fulbe clans of Hausaland in Nigeria's public

[180] https://wondertvmedia.com/hunters-arrest-five-man-kidnap-gang-in-kogi/

and governance structures. Multi-billion dollar arms deals for ranking Fulbe members, and multi-million dollar contracts to develop local alternatives to twitter and other curious awards in sectors running from Telecomms to the Oil industry gives credence to the plundering of the resources of indeginous countries and peoples through the instrument of the Nigerian State.

While it is difficult to fault anyone's quest for puritan living or religion, the outcomes can certainly be questioned, especially when these outcomes pose significant threats and potentially negative externalities to others in the community and those in adjoining countries who are legally and fiscally conjoined through the Berlin Conference.

So, one man's defense of Hausa country from the Kanuri looks very much like a launchpad for Southward invasions and attempts to conquer and establish a foreign ruling class over indigenous countries through means that universally qualify as terrorism.

This has aggressively intensified with the Buhari administration, and there seems to be a plan to cover it all up. Under foreign (possibly the US) pressure, some "Nigerians" were identified as sponsors of terrorism, most likely with funds plundered from Nigeria's Central Government and agencies. It followed the usual pattern when the Government suddenly announced that it wasn't interested in naming and shaming these sponsors of terrorism, but was interested in bringing them to book.[181]

It is clearly a valid concern how a Government that is more concerned about not shaming terrorists, will confidently

---

[181] https://www.premiumtimesng.com/news/top-news/485917-nigeria-not-interested-in-naming-shaming-alleged-terrorism-sponsors-presidency.html

prosecute them. President Buhari's Attorney General seems to have given an indication to this riddle; Pardon them. [182]

The strategy is very simple. Take control of a community or people, plunder the resources, and use the same resources to implement an agenda that is nowhere near Godly, as claimed.

Most of the peoples and countries in the Nigerian State have no problem with the Fuduyan Fulbe control or defense of Hausaland, and the interaction this causes with their Kanuri frenemies. This interaction, is responsible for almost all the terrorism plaguing the Nigerian State and has produced two of the four most dangerous terrorist groups in the world. Kanuri country can now list the North East Development Commission, with a first line charge from Nigeria's Federal budget, as a proceed of the terrorism.

Military officers have tried and died in the attempt to cut off Kanuri and Hausa country from the rest of the Nigerian State. Sadly the poor and impoverished peoples of these countries get blamed for the misdeeds of a few leaders, enablers and captors.

The question facing Nigerians (especially the leaders of thought) is 'what do you do about Hausaland'? There is a clear mass Stockholm Syndrome situation, and many of the 'traded souls' are eager to fight and kill at the instruction of their Fulbe defenders.

On the one hand, ignoring the fire in your neighbor's house means that it could soon spread to yours. If Gobir did not happen, Abiodun would not have given the innocent looking strangers permission to graze South of Jebba, and Ilorin would not have happened.

---

182 https://www.sunnewsonline.com/weve-received-320-applicants-from-convicts-seeking-pardon-malami/

On the other hand, some believe that the situation in Hausaland has gone beyond what any external intervention can salvage, and with the foreign Fulbe defenders of Hausa Country now switched to full attack mode on the rest of the Nigerian countries with NIN, Ruga, Farm Settlements, Herdsmen, Boko Haram etc.

It may not be the right time to worry about your neighbor, but instead it may be time to close your own gates and ramp up your Defense.

Using the light-hearted example of the FIFA World Cup, it is noteworthy that despite many countries spending millions of dollars to recruit foreign coaches for their national teams, foreign coaches have won the world cup exactly zero times of the 21 times that the FIFA World cup has held, and indigenous coaches of their national teams have won it all 21 times[183].

There is something that goes beyond the defense of the land of subjects and extremism that the love of country means to the indigenous peoples of a country. God knows that and that is why He strongly warned against countries having foreign rule and rulers.

---

183 (Wikpedia, 2021)

Tiv traditional rulers, to the South of Hausa Country, raise alarm of attacks by Fulbe Militia, similar attacks in Southern Kaduna, Jos Plateau and other parts of Nigeria[184]

INFORMATION, HISTORY AND CULTURE

With the very real and shocking struggles the Nigerian State is facing (hosting some of the most dangerous terrorist groups, most out of school children, highest number of poor people, a few extremist foreign clans having a tight and crazy grip on the Nation), many Nigerians, within and outside the country are beginning to question and investigate the sorry state of things, especially since Nigeria is made up of sustainable, and even wealthy indigenous countries.

---

[184] https://dailypost.ng/2021/09/28/benue-herdsmen-targeting-us-for-elimination-tiv-traditional-rulers-raise-alarm/

Regarding the indigenous countries, all kinds of disingenuous numbers are maliciously and ignorantly bandied about, suggesting that Nigeria is made up of so many ethnic groups, that it could never work as separate legal and fiscal units. This is obviously not true.

There is a part of Nigeria (stretching across the border in Cameroon) that is a true global phenomenon, that reportedly accounts for 10% of all the languages in the world. Over a hundred of these languages (and possibly up to 200) have less than 5,000 speakers. This is the source of the so-called 300 or so ethnic groups. An aberration in a corner of the country.

Modibbo Adama was able to take advantage of this fact to unleash killings and plunder in this part of Nigeria, and he got a state named after him as a reward. Current (as at 2021) 'Adamawa' (Gongola works as well) Governor, Umaru Fintiri, who rose through the political and legislative ranks to become the Speaker of the State, has done a tremendous job of trying to create a common cause for peoples previously seen as disparate, but it remains to be seen if he survives the assault of those who see his emergence as a threat to the owners (or defenders) of the area and Northern Nigeria in general.

After accounting for the 'aberration' around Gongola, Nigeria's ethnic countries and nationalities are actually very simple and straightforward and can be easily organized as a Confederation of Nigerian States, that will be largely autonomous legal and fiscal entities, with separate constitutions, but remitting residual funds to the centre for very few functions, including the Military. These states will range from between six and eight, depending on a few factors (whether Kanuri and the Fulbe 'defended' Hausa country decide to blend as one for instance).

This will encourage all societies and constituent countries to quickly work towards sustainability, and will automatically end most of the terrorism and security issues.

There are enduring traditional and cultural institutions, some of which trace their history back to almost 1000 years, but which have sadly been attacked and somewhat eroded by Colonialism, invading foreigners, and the successor Nigerian State.

The Omo n'Oba N'Edo, one of the 'related' indigenous cardinal rulers in the current Midwest and South Western areas of Nigeria got a rude awakening in 2021 with some foreign persons testing the waters in his domain[185].

The Alaafin, whose 200-year travail is well documented, still sits in a 'temporary' capital 100 miles south of (Old) Oyo. He is sometimes locked in a supremacy battle with the Ooni. The Ooni and the Omo n'Oba N'Edo seem to be in a weird 'Whose your Daddy' battle (Oduduwa does sound like Ereduwa), and the Attah Igala has thrown his hat into the ring by claiming to be everybody's Daddy.[186]

Little is heard from the Alaketu, but perhaps that is because his town was drawn off from most most of his country by the Berlin 'border drawers', so what most people in Nigeria (especially Lagos) know of Ketu (or Ketou) is a big market, close to Mile-12 in Lagos. Other towns like Sabe, Popo and Ajase (Porto Novo) were also randomly cut off.

Sitting just below (or beside) these Cardinal traditional rulers are between 20 - 30 principal rulers, who have command of

---

185 https://www.vanguardngr.com/2021/08/outrage-over-planned-turban-of-sultan-of-shuwa-arab-in-edo/

186 https://www.facebook.com/187757678021480/posts/igalayoruba-relationshipthe-oduduwa-conceptionthere-is-no-doubt-whatsoever-that-/842501302547111/

one to two million people and the large towns and the surrounding villages.

It is sad that many young people have little opportunities for interaction and engagement with the traditional stools they know they are meant to honour and respect, at least on the Western end of the Nigerian State. Other parts of Nigeria are fiercely republican.

Many people only see these leaders when they are stepping out with wives, or announcing the birth of new babies.

These leaders need to remember that similar bickering led their progenitors to see Ilorin as an 'Oyo matter', with only the Borgu princes trying to rally some sense into their bickering Southern neighbours. Eventually Oyo fell, and hundreds of thousands of refugees poured Southwards, and the country became open to Mohammed Bello's slave raiders.

Some kings including the Ooni of Ife (Modakeke) and the Owa Obokun of Ijesha (Osogbo) had to create new towns beside their kingdoms to accommodate the refugees. Other new towns like Ibadan evolved due to this event.

This was no longer an Oyo matter, and it was almost too late when a subsection of the kings got together to push the 'defenders and defended' of Hausa country out of their lands at a battle in Osogbo.

An interesting event played out with the captives of that battle. The foreigners were released with a message to their leaders that could be paraphrased as '*We do not know you, and have no business with you, stop invading our villages and killing our people*'. Those deemed not to be foreigners did not get that privilege.

These traditional rulers, though not having the far-reaching powers they once did, can still play subtle, but impactful roles, as their people face all kinds of attacks, and desperately look somewhere for leadership and direction.

Historically, traditional rulers also played key economic roles, by controlling commerce and markets for instance. With alleged guises of invasion through RUGA, Farms and livestock transformation, the traditional rulers can request possession of any lands and funds created for that purpose, and train (coordinate programs for) indigenes in the livestock supply chain.

If this and other creative initiatives can be birthed, the people can finally start to find their bearing, and engage with their roots, while solving a security and economic problem.

There is no need for AK-47 wielding drifters, whose well-watered country is over 1000 KM away to be planted within communities that are already being attacked by the same folks.

COMMENTARY 4 - A PRESIDENT'S CONCERN FOR HIS HOMELAND

The table below, attempts to summarise the publicly available information on the 'END SARS' protests in Lagos, and the events surrounding the controversial clearance of protesters at the Lekki tollgate, which many observers say was accompanied by the shootings.

	Event(s)	Independent sources	Comments
1	Following an online campaign that had been on for a few weeks, A-list musicians Runtown & Falz joined youth activists In an 'END-SARS' protests against Police brutality and bad governance in the country in the Lekki area of Lagos.	https://silverbirdtv.com/featured/74632/entertainers-falz-runtown-lead-endsars-protest-in-lagos/ https://businessday.ng/security/article/falz-runtown-join-forces-with-youths-in-endsars-protest/	The protests were peaceful and remained peaceful even as hired thugs tried to attack. Other Celebrities spotted at the protest in the following days

2	The Protest soon spread to many states in the Country, and Nigerian embassies abroad. Protests defy threats and reported attempts to bribe 'organizers'.	https://www.dw.com/en/opinion-nigerias-endsars-protesters-need-a-leader/a-55325920	Most people clearly identify with the stated cause of the protest, but informed minds also worry about the strategy of the well-coordinated, but seemingly leaderless movement. A strategic retreat is suggested by some. Calls for the President to address the county on the protests (then in its 3rd week) reach a fever pitch but are hardly acknowledged by the Presidency.
3	Oct 20 - Lagos Governor announces a Curfew a few hours before take off. Lagos thrown into confusion. Most hear less than 2 hours to curfew. Rumors later emerge about Cameras being removed from the Toll gate	https://www.garda.com/crisis24/news-alerts/391106/nigeria-24-hour-curfew-declared-in-lagos-state-amid-endsars-protests-october-20-update-13 https://edition.cnn.com/2020/10/20/africa/nigeria-lagos-curfew-intl/index.html https://news.trust.org/item/20201020103418-rsr2a	This is immediately interpreted as external pressure coming to bear on the Lagos Government. With the eye of the press on the Lekki toll gate protesters, this could be an opportunity for the police to pretend to show some competence by using water cannons & maybe 100 policemen in proper riot gear (a real kits challenge), even if only to window dress for the international media. No such luck!

4	Reports of killings surface on social media about 8pm.	https://www.channelstv.com/2020/11/21/lekki-shooting-army-admits-live-blank-bullets-were-taken-to-protest-venue/ https://www.premiumtimesng.com/news/top-news/425228-lekki-shooting-we-fired-blank-ammunition-army-tells-lagos-panel.html http://saharareporters.com/2020/10/25/revealed-nigerian-soldiers-involved-lekki-massacre-took-eight-bodies-bonny-camp-clinic	Pulling the pieces together, it appears Brig. Gen Francis Omata, Commander of the 81 Garrison Div, VI which is less than 5Km from the toll gate hears about it & takes off for the scene. Screams to identify himself by name. It appears operation was led by a Lt. Col S. O Bello. No clear answer on who authorized the operation even after a few months.
5	4-5 days of the worst looting and carnage Lagos has ever witnessed follows the shooting. Hardly any police or security presence anywhere	https://www.ft.com/content/04ab5d04-3647-4768-acea-7bdd47740c35 https://www.premiumtimesng.com/news/headlines/422994-looting-across-nigeria-as-arsonists-hoodlums-thieves-take-control.html https://www.bbc.com/news/world-africa-54633337	It is not uncommon to see up to 8 different security agencies harassing motorists on a typical Lagos day. However, none of these agencies, or the others like the OPC, Amotekun, Neighbourhood Safety Corps show any presence in addressing the carnage. A sad commentary of the lack of law enforcement (but plenty of bribery and harassment) in the county.

		https://www.thenationalnews.com/world/africa/nigeria-shooting-fans-long-simmering-anger-at-the-state-1.1098037	
6	Crazy claims and reversals. Up to 7 versions and reversals of events by the Government and Security agencies	https://apnews.com/article/africa-nigeria-shootings-military-intelligence-lagos-5b386c793c14ecf139d882d7ca4798f8 https://www.premiumtimesng.com/news/headlines/422021-endsars-despite-contrary-evidence-nigerian-army-denies-shooting-at-protesters.html https://www.vanguardngr.com/2020/11/still-on-lekki-shootings/ https://www.news24.com/news24/africa/news/nigerian-minister-says-its-unclear-who-shot-protesters-in-lagos-20201102	Governor says forces beyond his control responsible, Army says the Government invited them, after initially denying being at the tollgate. Governor denies inviting the Army; Army says they were there, but did not shoot; then we shot but only used blanks; and then we used some live Ammo but shot upwards.

7	CNN & Fisayo Soyombo (a leading and trusted investigative journalist) confirm killings. Maybe over 20 victims. Amnesty International reports much higher fatalities from the shootings and the following violence	https://www.amnesty.org/en/get-involved/take-action/nigeria-end-impunity-for-police-brutality-end-sars/	
8	Reports of harassment and intimidation of protest leaders and family members of victims	https://www.reuters.com/article/nigeria-protests-aftermath-insight-idUSKBN28A19W	
9	Nigeria's Federal Government steps up harassment and witch-hunting of protest organisers and sympathisers	https://www.hrw.org/news/2020/11/13/nigeria-punitive-financial-moves-against-protesters https://edition.cnn.com/2021/02/11/africa/nigeria-protests-acounts-unblocked-intl/index.html https://www.voanews.com/africa/6-nigerians-sentenced-funding-boko-haram-terrorist-group	Blocking of bank accounts, bank account blockages and other harassment measures. Quite interesting that the Government failed to block accounts of those linked to funding terrorist activities

10	On February 6 2021, the Lagos Panel of Enquiry gives the Lekki Concession company the right to reopen the toll gate.	https://www.aljazeera.com/news/2021/2/11/nigeria-government-warns-lagos-protesters-against-endsars-rally https://www.premiumtimesng.com/regional/ssouth-west/441027-breaking-lagos-judicial-panel-okays-lekki-toll-gate-reopening.html	Somewhat symbolic the 'Baba Kekere' Lateef Jakande, journalist, Action Group pioneer (jailed with Awolowo and Enahoro in the 60's) respected and high achieving former Governor of Lagos, whose administration built the VI, Lekki, Epe road as a free, non tolled road for Lagosians, died in the same week

As the End-SARS protests dragged on to the third week on October, 2020, there were calls from several quarters for Nigeria's President to address the protestors, or the nation, even if only as a symbolic show of empathy and understanding of the issues causing so much anger about the state of affairs in the country. President Buhari's actions suggest he had other things on his mind. On October 15, 2020, he hosted fellow Fulbe leaders Macky Fall of Senegal and Umaro Embalo of Guinea Bissau (Gambia's Adama Barrow was missing, but given Senegal's literal and figurative coverage of Gambia, may be assumed to have been informed of the meeting's outcome).

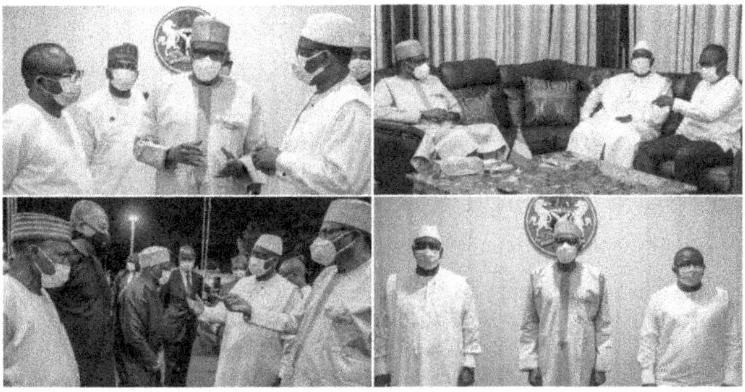

President Buhari on Thursday, October 15, received the Fulbe leaders of Guinea Bissau and Senegal on the eve of the elections in Guinea

The agenda of the closed meeting was assumed to be the Presidential election in Guinea which was held two days later, on October 17. Guinea is the heart of Fulbe country, but is yet to have a Fulbe leader, despite the Fulbes being the largest ethnic group in the country. In the past, all the other ethnic groups have combined against the Fulbe during elections, which has been quite embarrassing. There are strong indications that the Fulbe clans that run Nigeria decided to take the lead and put an end to this.

Buhari, with Katsina's Fulbe Governor Masari, hosts Conde in Daura.

Buhari, and later Ganduje (believed to be ascending close to the zenith of the informal[187] Fulbe hierarchy in Nigeria, possibly resulting in clashes with El-Rufai and Lamido Sanusi) had treated Alpha Conde to grand Durbars in Daura and Kano in 2019 and it was thought that they had received undertakings that Conde would not seek a third term and form an alliance to help the smooth emergence of the Fulbe candidate in Guinea.

Conde with Putin. Following Conde's overthrow in Guinea (led by the French trained Head of his Guards Unit in September 2021), France has reacted aggressively to Mali trying to get Russian help to solve its terrorism problem.

Whether it was a case of the old fox running a game on West Africa's ultimate game masters or it was just a case of Russia trying to protect its interest in Guinea's bauxite, Conde 3rd term was looking more expedient for Conde and Russia, and whatever commitment Nigeria's Fulbe clans had secured for their homeland was thrown out of the window.

[187] The formal structure is quite defined with Sokoto at the Zenith, but individuals outside the structure (even if members of the clan), sometimes complement, and at other times challenge the structure. (Gumi and Ahmadu Bello were examples of challengers of the structure)

Ganduje and Co. roll out the red carpet for Conde in Kano.

Not one to take real or perceived humiliation kindly, Buhari helped force the emergence of a Fulbe candidate in the smaller Guinea (Guinea Bissau) in January 2020 and Embalo seemingly, wholly brought into the "Bye Bye Conde, welcome Fulbe rule" Guinea Project.

Embolo (M), on one of his many shuttles to Abuja, as the Guinea elections approached.

He made at least five known visits to Buhari in 2020, including one supposedly on an errand to personally deliver a Covid-19 herbal solution from Madagascar (you can't come up with funnier content, even if it was fiction).

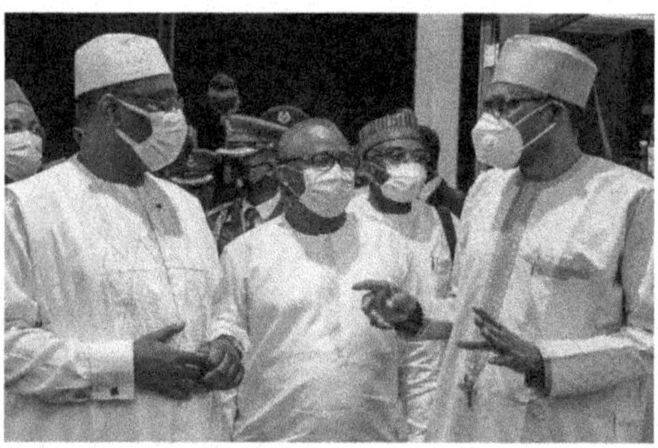

From left: President Macky Sall of Senegal; President Muhammadu Buhari and President Umaro Sissoco Embaló of Guinea Bissau during a dinner in honour of the visiting Presidents at the Presidential Villa in Abuja on Thursday.

It is believed that all the planning came to a head with the Abuja meeting of West Africa Fulbe heads of state about the election in the homeland on October 15, 2020.

It is not known if the quickly contained insurrection in Conakry on the eve of the election had any links to the Abuja meetings, but it can now be clearly seen why the End-SARS protests in Lekki was barely on the radar of Nigeria's president. He was focused on events in his homeland.

Additional References

https://www.vanguardngr.com/2020/10/guinea-president-names-street-after-buhari-for-his-fear-of-god/

https://www.theguardian.com/world/2019/aug/27/russians-have-special-status-politics-and-mining-mix-in-guinea

https://www.premiumtimesng.com/news/top-news/346096-guineas-alfa-conde-meets-buhari-in-daura.html

https://www.barrons.com/news/nigerian-leader-gets-samples-of-madagascar-virus-potion-01589650805

https://www.news24.com/news24/africa/news/mali-approaches-russian-private-companies-slams-france-at-un-20210926

GODLY GOVERNANCE, SOCIETIES AND NATIONS

This section navigates God and society, since a lot of the issues and controversy in Nigeria and elsewhere revolve around the quest (by indigenes and foreigners) to either establish Godly and sometimes puritan religious states or communities, or guide existing societies with Godly principles.

Apologies to readers who do not have a belief in God or do not believe that Godly principles lead to desired outcomes for societies. Apologies also to those with alternative or very different beliefs about God, His person, and what He represents.

This is not an attempt at cultural or religious imperialism, but underlying some of the more severe issues that have faced the indegeious peoples and countries of Nigeria for a couple of centuries is the claim that the enslavement and sale of the people, restrictions, destruction, killings, political systems, religious systems, overthrow of rulers, and the general quest for control over the people, their resources and societies is in order to ensure a more Godly society and country.

Questions regarding whether "Godly societies" translate to or mean "better societies," or are even intended to be, are probably beyond what this commentary on the Nigerian state should focus on, but Wole Soyinka and Ali Mazrui,[188] and some other brilliant minds[189] had some interesting exchanges that touched on this matter, that the intellectually curious can follow up on[190]. The exchanges mostly stemmed from the BBC/PBS series, The Africans: A triple heritage (that aired on NTA and was watched by author as a pre-teen) which was narrated by, and believed to be largely or wholly written by Mazrui.

The commentary here takes a much simpler approach. What has God revealed about Himself and his desire for societies in a way that can be validated and easily understood by just about anyone?

Scriptural quotes in this section will be from the Tanakh (accepted as scriptural by Jews, Christians and Moslems) and Gospels (accepted as scripture by Moslems and Christians).

LIES, THEFT, KILLINGS AND DESTRUCTION – THE WORKS OF THE DEVIL

Jesus (also called Yeshua (Joshua), Yasu, Jesu), the itinerant first century Jewish preacher in Roman controlled Galilee, who described himself as God's messiah, among other things, was remarkable in many regards. One of these was in an audacious, almost personal description and knowledge of the devil (Satan, Lucifer), a mythical being who is seen as the source and personification of evil in many cultures and religions.

This description is vividly captured in the conversation below, as Jesus began to draw the ire of not just the Jewish religious establishment, but of many regular folks in first century Judea.

---

188 https://www.africaknowledgeproject.org/index.php/war/article/view/422

189 https://www.jstor.org/stable/41069791

190 https://www.jstor.org/stable/2934911

Picking up the story from John[191]

> *"Abraham[192] is our father," they answered.*
>
> *"If you were Abraham's children," said Jesus, "then you would do what Abraham did. As it is, you are looking for a way to kill me, a man who has told you the truth that I heard from God. Abraham did not do such things. You are doing the works of your own father."*
>
> *"We are not illegitimate children," they protested. "The only Father we have is God himself."*
>
> *Jesus said to them, "If God were your Father, you would love me, for I have come here from God. I have not come on my own; God sent me. Why is my language not clear to you? Because you are unable to hear what I say. You belong to your father, the devil, and you want to carry out your father's desires. He was a murderer from the beginning, not holding to the truth, for there is no truth in him. When he lies, he speaks his native language, for he is a liar and the father of lies. Yet because I tell the truth, you do not believe me! Can any of you prove me guilty of sin? If I am telling the truth, why don't you believe me? Whoever belongs*

---

191 John 8: 39-47

192 Abraham, also called the 'Father of Faith' is a revered figure in Islam, Judaism and Christianity

> *to God hears what God says. The reason you do not hear is that you do not belong to God."*

Shortly afterwards[193], Jesus added

> *The thief does not come except to steal, and to kill, and to destroy. I have come that they may have life, and that they may have it more abundantly.*

The belief of the existence of a God or the Devil does not excuse or absolve anyone of individual responsibility for actions, and due apologies are offered to anyone who finds the suggestion of a God or anything beyond the material and temporal world, ridiculous, or unbelievable.

However, entertaining the possibility of a God, who is not only good, but has set standards for a good, prosperous, just and sustainable society definitely requires some examination, especially in the examination of the Nigerian space, where the mention of God litters private and public discourse, communication and meetings.

GODLY QUALITIES FOR SOCIETIES AND NATIONS.

Quality 1 - Love

(Ex 34: 6-7) [194]God describes Himself as the 'compassionate and gracious God, slow to anger, abounding in love and faithfulness, maintaining love to thousands, and forgiving

[193] John 10:10

[194] (Ex 34: 6-7) from the Jewish Tanakh, the source and foundation of the 'Abrahamic' religions

wickedness, rebellion and sin. It is also said that 'whoever does not love does not know God, because God is love'.[195] This can be very tough. Even when speaking about obvious evil and corruption, you are expected to somehow speak the truth in love. However, if God is love, then love must somehow be the most powerful force available. It is also very important to note that love is much more of a choice than an emotion. Love can also be said to be based on the value the 'lover' has placed on the 'beloved'. This is one of the primary driving forces of patriotism as the citizen places a greater value on the country (or Nation) than the reality of the situation may justify or demand. Among other attributes, love is said to (be):

a. Patient,
b. Kind,
c. Not envious,
d. Not boastful,
e. Not proud,
f. Not dishonouring of others, (so colonialism and shouting at 'Apes to Obey' for kicks were certainly not acts of love)
g. Not self-seeking,
h. Not easily angered,
i. Keep no record of wrongs (not sure if this means to forgive, or to be content with incomplete analysis. This book has obviously chosen the former interpretation. A proper understanding of History is essential to preventing recurrence of evil).
j. Not delight in evil but rejoice with the truth.

---

[195] (1 John 4:7-8) from the Epistles, which is only definitely considered scriptural by Christians

k. Always protects (Love for homeland should make one protect it from violent invaders and all forms of corruption and injustice)
l. Trust (The concept of fidelity is fundamental to good relationships, whether within a family, in business or society at large and is addressed in a few pages),
m. Hope,
n. Persevere.

Quality 2 - Truth

This also pertains to integrity. God does not lie or deceive.

God is not a man, that He should lie. [196]It is impossible for God to lie or deceive. Any system or Government based on lies or deceit has nothing to do with God. The truth is the truth. You may not know it in full, but certainly don't try to distort the bit you know. English politician Winston Churchill was quoted as saying 'The truth is incontrovertible. Malice may attack it, ignorance may deride it, but in the end, there it is'. Music Megastar Michael Jackson, expressed the same sentiment when he said 'Lies run sprints, but the truth runs Marathons'. This author has sadly experienced a few bold-faced high-stakes lies in his time, and sadly some of these relate to Nigeria, its structure and governance.

One incident that will linger for some time in the mind of this writer was NNPC's management confidently telling the President, Vice President, and other senior Nigerian government officials in October 2009 that they had fully participated and cooperated with the Finance Minister's team on the review of the Downstream Petroleum sector (Subsidies, Refineries, Petroleum Products) and were fully on board and accepted the

[196] Numbers

recommendations to end all petroleum (consumption) subsidies, sell all NNPC's refineries, concession the pipelines, and effectively end the corporation's involvement in the downstream petroleum sector.

This was in sharp contrast to the weeks of lack of cooperation, stonewalling and threats from people on the NNPC side, which continued in measured tones, as we stepped out of the Presidential meeting room. 12 years later, the refineries have not produced petroleum products, billions of dollars are reported as losses on refinery expenses almost every year, and an odious and senseless contract of $1.5billion has been signed to 'overhaul' one of the refineries.

In addition to the disregard for truth and a seeming disregard for information and intellect, there is a certain callousness towards the people in the way the affairs of the Nigerian nation is run.

Quality 3 - Righteousness and Justice

God is undeniably sovereign, but He strongly suggests that he has made himself bound to certain standards in His actions and dealings, especially with Humanity. *'Righteousness and justice are the foundation of your throne; love and faithfulness go before you'*[197] So, the sovereign God suggests that he has credibility because He is just and righteous in His dealings with people. *But let justice roll on like a river, righteousness like a never-failing stream.* [198] *The LORD loves righteousness and justice; the earth is full of his*

[197] Amos 5:74

[198] Psalm 33:5

unfailing love[199]*. The righteous care about justice for the poor, but the wicked have no such concern*[200]*.*

In summary, **Justice** pertains to equity and fairness while Righteousness can be said to be having a good and acceptable moral code, that should make an individual live a positive, disciplined life which should naturally extend to that person being a blessing and positive influence on his or her family, neighborhood and community, and as much as possible and within his or her control being at peace with everyone. Both of these qualities should naturally lead to an active and activated concern about the weak and vulnerable in society including oppressed minorities, women, children, the elderly, the disabled, and other vulnerable groups and people.

AK47 and RPG wielding kidnappers and terrorists are however not members of a vulnerable group, and those mobilizing them for more attacks with hundreds of millions of naira from the taxes of hardworking and honest Nigerians, without asking questions about the young students that these criminals, terrorists and violent propagators have murdered certainly have no part with God, or His righteousness and Justice.

Inequity and injustice are sadly built into the Nigerian state. From the violent propagations and relentless slave raids of the invading Fulbe clans spearheaded by Fuduye, Mohammed Bello and Modibo Adama, the associated killings of peoples who did not even know them, not to talk of wronging them in any way, to the formal English colonialism, to the effective continuous rule by foreigners over native indigenes through violence and flawed colonial and national structures.

199 Proverbs 29:7

200 Psalm 89:14

The brazen theft of resources continues largely to sustain these foreign interests and establishments that claim a dubious legitimacy through the mess of a structure that is the Nigerian state.

God is righteous and just in His dealings with mankind. No righteous person will initiate or perpetuate injustice, and justice is rooted in righteousness.

Quality 4 - Work Ethic

No nation can be sustainable and prosperous in the long run without a good and positive work ethic. Despite remaining as the indisputable leader in knowledge and research globally, one of the reasons the USA has struggled lately is because a growing number of people want to have other people do (most or all of) their work but make the same amount of money; the honest month's wage' from the 'honest day's job mentality).

This also extends to the 'Pharma Bro' strategy that is widely prevalent where there are continuous attempts to separate fair value[201] of a service or product from cost or payment received. This also explains why the United States, despite being the leading research nation by far, has in recent decades, had some trouble translating the results of its own research into benefits for its own society ahead of other countries. Combining this with the impact of special interests, lobbying and the power of corporations, it can be seen why some medicines and medical procedures cost five times more in America, compared to other countries, or why the United States lagged Europe and parts of Asia in the widespread adoption of mobile telephones and technologies,

201 Measurements of fair value may differ, but there will always be a rational and decent range beyond which pricing is predatory.

despite the fact that the research and product development in some of these cases were fully done in the United States.

If a declining work ethnic can chip away at American prosperity, imagine the impact it has had in Nigeria, where the ruling Fulbe clans have destroyed the productive capacity of Hausaland (Nigeria's largest indigenous country), with an unnatural hegemony that is wholly focused on self-preservation of a tiny ruling class, which has stifled the possibility of a productive and decent life for majority in Hausaland. There was a fairly quick transition from taking over control of the wealthiest country in Africa at the time, to perhaps the largest and most coordinated commercial slave raiding and trading operations in the history of West Africa.

Going back to God and the account of Moses in the first book of his Pentateuch *The LORD God took the man and put him in the Garden of Eden to work it and take care of it.* [202] Shortly afterwards, a charge to subdue, govern, or take care of the earth is given to humanity. This suggests that gainful and productive employment has always been a fundamental part of mankind's existence on Earth, per the Mosaic account.

This writer would argue, with good evidence from God's revelation of Himself, that good honest work/labour is as much of an expression of fulfilling God's intention and desire for man's existence on earth, and maybe even more so, than some acts of worship dictated by religious institutions.

Any system that utilizes some supposed religious principle, basis or mantra to systematically transform the majority of the productive (male and female) base of society into an unemployed and unemployable legion that depends on aid and handouts is by

---

[202] Genesis 2:15

definition ungodly, no matter how observant and compliant with religious codes and regulation the system claims to be.

Welfare states, are only sustainable in the long run, when a significant majority (I'd say 90% - 95%) of the labour force of a society are productively engaged in gainful employment and making a decent income, stimulating economic activity, which naturally generates fair (activity based, primarily) taxes to be used by a small but effective government to carry out functions that the citizens are not ideally positioned to carry out.

Keynes, in his *General theory of Employment, Interest and Money*, reiterated the importance of people not getting paid for doing nothing, by suggesting that a man could be employed to dig a hole and another employed to cover it to ensure the productive base is not idle (we won't suggest the option of printing money and burying it to the current[203] 'Quantitative Easing' Central Bank).

Of course, digging holes should be far from the mind of any thinking government or administration in nation like Nigeria, with its huge infrastructure deficit. Thousands of men and women in each state and possibly district can be employed in the relatively simple process of constructing brick roads. With most of the country not on the electricity grid, hundreds in each locality could be trained and employed as solar installers, with regional governments investing in the supply chain infrastructure for solar infrastructure.

There are many other examples, but multiplier effects from these two simple options will probably be double digit GDP growth for over a decade. Road infrastructure has almost totally collapsed all over Nigeria, so a temporary (for 5 – 10 years) stopgap of bricks for maybe even Trunk-B road systems based

[203] 2021

on bricks is certainly not crazy, especially since brick roads are cheaper, last longer, and require less maintenance than asphalt roads.[204]

These will lead to a massive productivity expansion, environmental benefits, a consumption boost, and cost reductions from a scaled solar led electricity generation.

However, it is becoming clear that there is a strong preference to keep the majority of the people in certain sections of the country poor, ignorant and dependent, as an engaged an open mind will naturally start questioning and unjust setup, run by "foreigners." Quite naturally, increasing income levels will lead to a lot more demands over rights and liberty, which the people who see themselves as the guardians of an oppressive system are terrified of.

Quality 5 - Fidelity

Fidelity (faithfulness, loyalty) will be looked at within the context of a family, which should be the building block for a stable society, and within the greater society itself.

Within the family, the foundation of stability and fidelity is a home-based on a marriage between a man and a woman, at least based on God's standards. This is naturally the bedrock for a stable society, and the breakdown of the marriage and home structure in a society usually precedes various forms of instability, including the terrorism that is rocking Hausa and Kanuri countries in current day (2021) Nigeria, or the higher poverty rates and lower education achievements of the black Diaspora[205].

---

[204] https://www.aggregateresearch.com/news/brick-streets-hold-up-better-than-asphalt/

[205] https://ifstudies.org/blog/family-breakdown-and-americas-welfare-system

The abundance of millions of kids, who do not live in stable, two-parent homes (much easier for step-parents or fathers with multiple wives to permanently send 5 year olds out of the house and leave them to their fate or to a brutalizing cleric), and are left to fend for themselves under perverted religious practices has proven to be an extremely fertile breeding nursery for clerics to raise armies of terrorists who have never had the opportunity of experiencing the love of a stable, two-parent home as God intended.

Any parent, who discards, his/her primary function of raising a child in a stable home, and send 5-6 year olds to a 'cleric' to join dozens or hundreds of similar kids who have to beg for meals and daily existence is not performing a religious duty, but is a willing participant in the satanic destruction of the family, and consequently in the perversion of Godly society.

So, Marriage, as the bedrock of a Godly society, is probably the most important contractual, legal, intimate and personal relationship that any man or woman is meant to get into, and it is intended to be till the end of the physical life of either party. (The marriage should not hasten, or lead to the death of either party, obviously).

To go off on a tangent a little, Jesus, while responding to another trick question on marriage, made it clear that marriage and sexual relationships are limited to earth, or the physical plane, so marriage or sexual arrangements do not extend to heaven, paradise or any afterlife reward systems.

In traditional African society, as in almost all societies throughout history, monogamy was by far the dominant form of marriage, with polygamy largely limited to royalty and well to do landowners. So to suggest that 'Polygamy is African' is not true.

Is polygamy wrong, and a sin then?

This is one of the rare situations where the answer is nearly a yes, but not definitely. Put in another way, for almost all people, and in almost every situation, it is a yes, but there are situations that could be termed as exceptions.

Let's go back to itinerant Jewish preacher, Jesus of Nazareth. While speaking about marriage, in response to a question about divorce, he said, 'Haven't you read the Scriptures?" Jesus replied, "They record that from the beginning 'God made them male and female.'[a]" 5 And he said, "'This explains why a man leaves his father and mother and is joined to his wife, and the two are united into one.'[b] 6 Since they are no longer two but one, let no one split apart what God has joined together."[206]

Almost nobody will be thinking of devotion and fidelity to his wife, as commanded by God, while making plans to get married to another woman. Also, because women have to bear additional physical and biological burdens (pregnancy, for instance), loyalty and faithfulness on the part of the man means that he has to stick to his wife, even when he realistically believes he has options. It is not loyalty or fidelity when you have no 'actionable' options.

A man who cannot maintain fidelity within a monogamy, cannot be expected to show fidelity with state resources. That is why Polygamists are meant to be well to do people of private means, and this 'means' cannot be expected to be from the public purse.

Nigeria, facing a population crisis, and now with more poor people than any other nation, cannot afford to have a policy that lags the Taliban on this[207]. Polygamists are meant to be Independently wealthy and not fund their preference from the

[206] Matthew 19: 4-6 New Testament Gospels

[207] https://www.bbc.co.uk/news/world-asia-55630097

public treasury and can still contribute to the public discourse and serve the country without holding public office. Even the Taliban of Afghanistan, who rule by fear and oppression, seem to be coming on board regarding this idea.[208]

There seemed to be a possibility for an exception with regards to God's insistence on monogamy for the Jews under Mosaic law, with the provision for a Levirate marriage, or a "guardian-redeemer" in situations where a man died and left a wife without any children. This was supposed to continue the dead man's lineage, and also had some benefit in terms of the protection of the widow. Both parties were however expected to agree to the arrangement and had the option to decline.

An example (based on ignorance) is also usually made of Abraham, not unexpectedly, as the one to whom the Abrahamic religions refer. According to the account of Moses in Genesis (presumably written under God's instruction), at no point was Abraham ever polygamous. He had a short dalliance with Hagar, his wife's employee, in violation of God's direct instruction regarding how his biological heir was to be born.

There are many moral and ethical problems that flow from an employer directing an employee to have sexual relations with her husband, and it is quite understandable why God would be unhappy with the situation.

Interestingly, this scenario is also similar to that of Adam in Eden, where despite God's clear instruction, Adam yielded to contrary information from his wife despite his own authority and clear direction on the matter. In Abraham and Sarah situation, Moses reports a divine visitation to Abraham[209], with a couple of

208 https://swordstoday.ie/afghan-taliban-leader-bans-polygamy-for-his-commanders/

209 Genesis 18

accompanying angels to remind Abraham and Sarah about his promise to them about their own biological child and for them to keep faith and keep trusting. This also signals God's strong interest in the institution of marriage, as He designed it to be, a monogamy.

Upon the death of Sarah, Abraham married another wife, Keturah (who some suggest was Hagar), but at no point throughout Abraham's life was he a polygamist. Moses certainly provided instruction for the regulation of polygamy, especially for the protection of the women caught up in it, but that was clearly not an approval.

Some have suggested that as with divorce, Moses allowed polygamy because of the stubborn hearts of men who would sometimes commit spousal murder to get around the restrictions from divorce and polygamy.

So, is polygamy a sin then? Well not quite, especially if in the unlikely scenario of the Guardian-Redeemer, but outside of this limited situation, God does not suggest, promote or justify polygamy. He did mention that Kings (and by extension, I guess the rich, famous and powerful) will be able to attract multiple women as we see today, and as we've seen throughout history.

This is however not an approval or accommodation, as God warns kings (& everyone in the list above) to limit the number of wives they will have, because even He knows human nature, but human nature and tendencies cannot be an excuse or abdication of personal responsibility.

While Royals, Sovereigns, the rich and powerful might seem to have this tendency (and some would argue and say allowance) for polygamy, God seems to suggest the other extreme for religious men, women and clerics. So Monks, Priests sometimes choose to forgo the institution of marriage and wholly and completely focus on Godly work to prevent family and other worldly matters from getting in the way of their work for God.

In the New Testament Epistles, which are not held up to the scriptural standard of the Gospels and Tanakh by some in the Abrahamic faiths, Paul implied a limited sanction for Christians that are polygamous, which is that they shall not hold church office, or actually work in church. So they are not models of righteousness, but cannot be stopped from being part of a Godly assembly.

So, while it is not stated verbatim that polygamy is a sin, God clearly states that His design of, and preference for marriage is monogamy, and anyone in a polygamous set up almost certainly shows that trust and fidelity was broken at some point.

As a strong signal towards fidelity and God's stated and desired position, especially given the societal and economic problems from an unchecked population growth and the breakdown in the family structure, laws or statutes must be enacted in Nigeria, or its successor states, to make sure that everyone in a polygamous arrangement (outside of the first wife, who is usually the only 'victim' in the situation) should not be eligible for any type of state benefits, including employment, political appointments, or running for political office (the latter two should involve more sacrifice than benefits anyway). The same should apply to any man or woman with children from more than 4 biological births/pregnancies.

The causes and symptoms of some of the serious problems facing the nation(s), including the impending population nightmare, are staring right at us, and we must not play the ostrich for false considerations. [210]

---

[210] This means any serving public servant that takes a second wife should resign immediately. There should also be fines and loss of some public

Sadly in Nigeria, prominent top public servants in all the arms of Government and Society who should be examples, of a disciplined and model home life, boast about having 25, 30 and up to 80 kids. This includes men holding some of the highest offices in the Nation. This behaviour is extremely irresponsible and damaging as polygamy is naturally tied to a breakdown of the family structure and less care for children at all levels, but especially at lower income levels.

Suggestion that polygamy enables well-to-do men help single women and widows is pretty baseless. 'Help' does not have to be tied to sex and marriage. If anything at all, it suggests taking advantage of vulnerable women on a permanent basis (or at least till the man's lust is done). This is by no means a one way street, as some women are naturally attracted to the immediate comfort, safety and security that comes with a marriage to a wealthy polygamist. God clearly communicated that men with means and opportunity will take advantage of the situation, without (and many times, even with) the necessary checks and balances. Just about every man, and adult is well aware of this basic fact, with or without a divine revelation or instruction.

Women, having no intellectual distinction from men, usually have to deal with the biological burden of childbearing and a bigger share of child-raising and homebuilding responsibilities. While not getting into the science and debate around the social or emotional dynamics, it is obvious that a man needs to be a lot more devoted (show more fidelity) to a woman, given the relative ease with which he can wander off, since he is sexually available, and active for a much longer period on average. Perhaps that is why God specifically commanded a man to leave his past (relatively speaking) and be united to his wife in a monogamous

---

privileges starting from the time a man has his 5th biological child relationship since he is the one who needs a lot more prompting to stay in a marriage, or to keep it as a God-defined monogamous marriage.

FIDELITY IN SOCIETY

Historically in many villages and towns in several parts of Nigeria, people did not need to watch over their wares all the time. Buyers simply picked up the items and left the money, and would shout to get the seller's attention if unsure of the price. Several examples of this historical and cultural fidelity remain, especially with the Esusu & Cooperative model. The Esusu model has a standing 'Alajo', a traditional banker, that is still common in many markets (the musician, Brymo has a song about Alajo Somolu), while the cooperative or contribution model is common in offices where a group of workers contribute money each month, and based on a rotation, one person gets the contribution in a month. This is often useful for asset acquisition, since Nigeria's financial sector is not designed to support consumers or the real sector. Fela Kuti also alludes to this traditional African fidelity, a very godly quality which often does not need a formal contract, in 'Underground System', his tribute to Burkinabe President, Thomas Sankara.

Fela Kuti, in 'Underground System' a tribute to the murdered Burkinabe President, Thomas Sankara, suggested that Sankara's murder was a betrayal by ECOWAS leaders. of the traditional trust that is prevalent in African societies, Ex-President Blaise Compaore was charged for Sankara's murder in April 2021.

The respect for contracts and agreements is fundamental for a well-functioning society, and before seeking redress and enforceability in courts and official quarters, the societal norms and culture should be the first and most efficient source and guardian of trust and fidelity.

Quality 6 - No Foreign Ruler

In just a few years, the Action Group built more schools in the Western Region of Nigeria, which had about a quarter of the population of the country, than the English colonials built in the entire country in about 60 years of rule.

> [15] *be sure to appoint over you a king the LORD your God chooses. He must be from among your fellow Israelites.* ***Do not place a foreigner over you****, one who is not an Israelite.* [16] *The king, moreover, must not acquire great numbers of horses for himself or make the people return to Egypt to get more of them, for the LORD has told you, "You are not to go back that way again."* [17] *He must not take many wives, or his heart will be led astray. He must not accumulate large amounts of silver and gold.*[211]

The Jewish Tanakh, which is considered scriptural by Moslems, Christians and Jews tells a story of two women who came to King Solomon 's court with a living baby and a dead baby, with a dispute over who was the mother of the living baby. I reckon you are familiar with the decision of Solomon in matter and the method used to come to that decision. A critical part of that method is the psychological truth that those who do not have a close affiliation (in this case biological) with a person or a situation find it easy to settle for a conclusion that involves the killing of the person, or a complete destruction of the object in contention (be it a property or nation). So, the mother of the dead baby realized she could use the threat of death of the living baby to her advantage as the mother of the living baby would rather let her have the baby than entertain a situation in which the baby would be killed.

Interestingly, it is likely that an alternate ending, with the other claimant getting the child, would have eventually led to a life lacking in love and care, and a sooner than expected death

[211] Deuteronomy 17:15-17

for the surviving baby, due to the bitterness, and 'wickedness' with which the unnatural mother would have raised the child.

Perhaps, there could be some insight from this about the fate of Hausa Country over the last two centuries, and more recently, the Nigerian State.

The major problem of the Nigerian state is that it is ruled (sometimes directly, sometimes implicitly, but always effectively) by foreign adventurists who have no loyalty to the constituent peoples and countries. The English colonialists effectively handed the country to a bunch of foreign, invading Fulbe clans who believe that English colonialism only halted the expansion of their century old colonialism of Hausaland to the rest of the current day Nigeria. This belief is obviously false and deluded, but people are guided by, and act upon their beliefs, and the clans have been actively seeking to actualize their beliefs by any and every means (the documents called constitutions, religion, herdsmen, terrorism, election annulments, NIN, demands for Ruga, Farming Settlements, effective control of the central revenue and security agencies including the Ports, NNPC, DSS, Defence Agencies etc).

In the weird twist of the Nigerian State, institutions like the 'Federal Character Commission', which aim to ensure there is a reasonable spread of appointees from all over the country (a sign of a sick and faltering nation on its own), end up deepening the hegemony of the Fulbe clans since unlike indigenous Nigerian ethnic groups that are mostly domiciled in their homelands in sections and areas of the country, the Fulbe clans have usurped the traditional and religious institutions in all of Hausaland and some adjoining areas.

So cases of almost certain nepotism are waived away as "meeting the federal character principle," since members of the few Fulbe clans may form the religious and traditional ruling class in states like Niger, Kano and Adamawa, which are technically in 'different geo-political zones'. This was the explanation given by the Police Affairs Minister in 2021 to address the worrying pattern of the leadership of Nigeria's security agencies, including those covered by his Ministry. It is just a continuation of the 'poison pill' of 'stability over fairness (aptitude, ability and qualification) introduced into the Nigerian State by Ahmadu Bello and Mohammed Ribadu (both chieftains of foreign Fulbe clans).

This life-sucking policy continues to find expression in policies like the 'quota system' for Common Entrance to Secondary Schools where 9 and 10 year old kids are introduced to the stark

common entrance cutoff mark for Zamfara is 4, while it is 139 for Anambra.[212][213] Many of the kids carry the bitterness and disappointment in the Nigerian State for life.

What type of nation deliberately sows anger, bitterness, and hatred for the country in the minds of 9, 10 and 11 year olds over the years through blatant injustice and disregard for competence, all in a bid to ensure the smooth continuation of a hegemony?

The answer is simple; a nation that was set up by foreigners, for the interest of foreigners, to be looted by foreigners, while entrapping the constituent indigenous countries and peoples in a dysfunctional and unworkable arrangement.

From the criminal gathering known as the Berlin conference to colonialism; to Lugard setting up the nation as two protectorates for a simpler administration that will accommodate his

212 https://servantboy.com/national-common-entrance-cut-off-mark/

213 https://businessday.ng/education/article/how-cut-off-marks-for-nigerias-unity-schools-may-be-destroying-quality/

interests in England; to the eventual absurd amalgamation that was sealed off by the removal (or flight in some cases), and then the reinstallation of the foreign Fulbe rulers of a Hausaland that had retrogressed from the 'richest country in Africa at the time of the Fulbe takeover to an unsustainable basket case (under the chokehold of a ruling class that despises its so-called subjects) that had to be propped up by other countries in the 'middle-belt' of the country; to the amalgamation with the 'Southern Countries' to cover the deficit of the now unsustainable & poverty stricken Hausa country that could not be covered by the gerrymandering of the 'middle-belt' to Hausa Country and the parts of Kauri Country in Nigeria.

And eventually to the post-colonial entity that has systematically taken the rights, resources and ability to function from the indigenous ethnic groups and countries and concentrated the control of the resources and security of the country in the clutches of invading religious clans who are presumed to be fighting a proxy war for full control of the country with some of most dangerous terrorist groups on the planet in tow[214].

The indigenous peoples and countries that constitute Nigeria have mostly been extremely accommodating towards immigrants and visitors, and remain so. However every country must determine the terms on which it accepts immigrants. Terrorists, violent propagators, slave raiders, bandits, colonialist, destroyers of standards and norms that have built and sustained societies for generations who have no historical precedent of building any decent or sustainable societies of theirs, cannot just stroll into countries that have existed before the Berlin conference and English colonialism and force their way into the Nigerian State,

---

214 https://guardian.ng/news/nigeria-retains-position-as-the-third-most-terrorised-country-in-the-world/

and its constituent countries and claim ownership and rule of the country on some ridiculous religious angle and basis.

Quality 7 - Freedom and Liberty

Jesus quoted Isaiah's prophecy about bringing freedom to the captives to begin his ministry and push his claim as God's messiah. Quite expectedly, the epistles, which explain how faith in Christ should be lived out, are littered with notes and concepts about freedom, liberty and redemption. However, this book will only quote from the Gospels and Tanakh, which also speak about freedom and liberty, beginning from Isaiah's prophecy, which Jesus claimed was fulfilled by him.
Isaiah's prophecy:

> *The Spirit of the Lord GOD is upon me,*
> *because the LORD has anointed me*
> *to bring good news to the poor;[a]*
> *he has sent me to bind up the brokenhearted,*
> *to proclaim liberty to the captives,*
> *and the opening of the prison to those who are bound*

And the original Dr. Luke's account of Jesus claiming the fulfillment of the prophecy[215]

He went to Nazareth, where he had been brought up, and on the Sabbath day he went into the synagogue, as was his custom. He stood up to read, [17] and the scroll of the prophet Isaiah was handed to him. Unrolling it, he found the place where it is written:

[215] Luke 4:18 - 22

"The Spirit of the Lord is on me,
because he has anointed me
to proclaim good news to the poor.
He has sent me to proclaim freedom for the prisoners
and recovery of sight for the blind,
to set the oppressed free,
to proclaim the year of the Lord's favor."[f]
Then he rolled up the scroll, gave it back to the attendant and sat down. The eyes of everyone in the synagogue were fastened on him. He began by saying to them, "Today this scripture is fulfilled in your hearing."
All spoke well of him and were amazed at the gracious words that came from his lips. "Isn't this Joseph's son?" they asked.

Jesus also clearly communicates the idea of liberty and an improved quality of life from faith in God and Godly societies.

So Jesus said to the Jews who had believed in him, "If you abide in my word, you are truly my disciples, and you will know the truth, and the truth will set you free."[216]

The thief comes only to steal and kill and destroy. I came that they may have life and have it abundantly[217].

In the Tanakh, Jeremiah, the lamenting prophet, speaks about God's anger towards slavery, and His desire to have those in any form of slavery freed. [218]

You recently repented and did what was right in my eyes by proclaiming liberty, each to his neighbor, and you made a covenant before me in the house that is called by my name, but then you turned around and profaned my name when each of you took back his male

---

216 John 8:31-32
217 John 10:10
218 Jeremiah 34: 15 -17

and female slaves, whom you had set free according to their desire, and you brought them into subjection to be your slaves. "Therefore, thus says the L*ORD: You have not obeyed me by proclaiming liberty, everyone to his brother and to his neighbor; behold, I proclaim to you liberty to the sword, to pestilence, and to famine, declares the* L*ORD. I will make you a horror to all the kingdoms of the earth.*

Putting Women in Their Place?

No society can lay a credible claim to being free or having liberty if half the population is subjugated, oppressed, or repressed.

One of the more remarkable aspects of the ministry of Jesus, especially to those observing it real time, was his interactions with women. We were told that his disciples sometimes wondered, almost disapprovingly, about his interaction with women. Why would this miracle-performing Rabbi, who claimed he had existed before Abraham, and had the authority to forgive sin, interact with women he was not 'related' to, especially since some of these women were not society's standards of decency and morality? Jesus even had the audacity to have women as disciples, 'apart from the twelve', including Mary Magdalene, Joanna, Suzanna, and other women who provided for Jesus' disciples 'out of their own resources'. Some have even suggested that the seventy pairs sent out by Jesus on the first known evangelical mission were pairs of men and women, which would have drawn the ire of the religious establishment.

The rights of women in first century Israel had become severely limited as the Talmud, the collection of oral interpretations and applications of the Torah (Jewish law) by Rabbis over the previous hundreds of years, governed daily social, cultural and to some extent, economic life.

Women were not expected to work outside the home, and even sometimes needed a male relative to accompany them on routine tasks like going to the market, similar to life in some of today's extremely restricted societies. This was a complete turnaround from much earlier in Israel's history when Deborah served as the Nation's leader (Judge) before it had a king.

It is therefore not surprising that the religious establishment was looking for every opportunity to get rid of this miracle-performing rabbi, who constantly reminded them that their fasts, religious dressing, food abstinences and other religious codes meant nothing, if their hearts were not right. It amounted to 'meeting the letter of the law, and not the spirit'; Hypocrisy; and having a form of Godliness, but denying the power therein (of God). Their harsh interpretations of what a Godly society should be, was actually making life difficult for the people, and putting barriers between the people and God, the exact opposite of what Godly societies are meant to do, according to Jesus, who compressed the most important commandments as 'Love God, Love people'

Jesus clearly knew that the subjugation of half of society through selfish and egocentric societal or religious codes was not the will of God, and he directly confronted this evil, to the chagrin of the religious order of the day, and his own male disciples (and possibly a female one, as Martha wanted her sister to join her in the kitchen rather than sit at the feet of Jesus with other guests).

Thankfully, suppression of women is alien to almost all indigenous cultures in Nigeria, including Hausaland (historically), where the story of Queen Amina, who led her people in battle, and at peacetime, is legendary. Many cultures have had women as regents, priestesses and permanent members of the Kings' Councils. Women have also been at the forefront of economic activity, playing dominant roles in farming, pottery, trading and weaving.

Many have insightfully suggested that widespread oppression of women usually corresponds to a strong satanic influence in a society, noting that a long standing enmity exists between Satan and women from the beginning of humanity according to Moses in Genesis.[219] The writer observed that women tend to form a significant majority in religious congregations where there are no restrictions. Women also tend to be more reliable and responsible in the home, where they are able to stretch resources, and make the most with very little.

sonal observation) who don't mind making a deal with the devil for '15 minutes of fame, wealth or power'. For those resident in Nigeria, images of young men bathing in cemeteries, undergoing crazy rituals and eating things that should not even be repeated, have made the rounds in recent years.

Stepping aside from personal observation, empirical evidence clearly shows that women's education and empowerment is one of the most powerful drivers of social and economic development. [220]

No society can make any significant progress by holding down (many would argue the better) half of its brainpower and human resource. God knows that, so it is quite strange why men (males, this time) with slight advantages with brute strength and biology (not burdened with pregnancy etc) feel that they can use 'devotion to God' to hold women down. Any society that

[219] Genesis 3:15

[220] Low Schooling for Girls, Slower Growth for All? Cross-Country Evidence on the Effect of Gender Inequality in Education on Economic Development. World Bank Economic Review, Dec. 2002 https://doi.org/10.1093/wber/lhf004

allows this usually pays a heavy price, as it ends up as a poorer and weaker society.

Godly Nation, or Godless Contraption?

It is important to note that the Nigerian nation is radically different from most of its indigenous constituent countries. The invading Fulbe clans, have however (with a religious hook and covert English help, most would agree), tightened their grip on the largest indigenous ethnic group, the Hausa, and have aggressively tried to use that as a launchpad to rule the rest of the country, and critically, control the oil of the Niger-Delta, which some suggest has been an important secondary target for the last 50 years.

So, in summary, Nigeria was:

- Conceived in corruption (Berlin)
- Birthed in corruption, deception and violence (English, Lugard)
- Exists for corruption on the basis of 'injustice and designed ignorance' of the majority, and is
- Run by corruption and deceit.

As a matter of fact, some suggest that when Lugard's occultist leanings and frenzied detour to upstage and reinstall the Fulbe rulers of Hausaland are juxtaposed with Fuduye's 100 year prediction of his lineage's reign on Hausaland until its foreign conqueror showed up.[221] Lugard's strange actions that eventually led to the amalgamation of 1914, has led some to suggest that the realm of Hausaland's Fulbe rulers now extends to all of Nigeria,

[221] Waiting for the fulfillment of prophecy in Nigeria – New York Times, August 11, 1912

initially in a secondary capacity, under British Colonialism, but in a primary, even if subtle form, thereafter.

Of course, many would disagree that physical, temporal events are influenced or controlled by spiritual forces, and regardless of your thoughts on that, what cannot be disputed is that the slave-raiding Fulbe clan leaders knew that Lugard had done them an unbelievable and unmerited favor. They certainly responded to their overthrower and restorer with a lot of adulation and respect.

Lugard and his Fulbe Clanheads.

So, are Nigeria's foundations, structure, and administration based on or influenced by Godly principles, godless ones, or something more sinister?

It should not be too difficult for honest and rational minds to come to a conclusion on this.

COMMENTARY 5 - BLAME IT ON NIGER

B*lame it On the Rain,* the song (or lip-sync) from the '80s and 90's German-French duo Milli Vanilli must be one of the favorites of the Fulbe overlords that run Hausaland, and who seem to be on a mindless quest to extend that control (through Ruga, NIN, Farm settlements etc) to the rest of the indigenous countries that are partly or wholly trapped in the current Nigerian State.

You just need to swap 'the rain' for 'Niger' in the song. Any time Nigeria's foreign Fulbe overlords are caught in an embarrassing situation that should highlight just how much devastation their rule has done to Hausa Country over the last 220 years, they're quick to say, 'well, blame it on Niger.' They play on the ignorance of most, especially in the middle belt and south of the Nigerian State, who do not know that Niger is primarily and predominantly the northern half of Hausaland, with about 55% of the population of the country being Hausa. The other major ethnic group with between 20% to 25% of the people are the Djeram/Songhai, who occupy the southwest of the country and have (in)famously had historical interactions with the Borgu kingdom. With the sprinkling of the Tuaregs, Kanuris, (Shuwa) Arabs, and the Fulbes who have become the ruling class of Hausaland on both sides of the border, and you have Berlin created the Niger Republic.

So there's no difference between the northernmost parts of Nigeria which is Hausa land and Kanuri land and the southern part of Niger, which is also Hausa land and Kanuri land. The criminals gathered in Berlin to draft initial maps for Africa split Hausa Country into two halves of British northern Nigeria and the French Niger. Nigerians Fulbe Overloads know that. But most Nigerians do not know or care since they're mostly concerned about their own indigenous countries that have been consumed into this colonial mess.

It appears Niger has also recently become an easy way to hide the proceeds of some of the massive corruption perpetrated by the overlords of Nigeria. It is an easy and convenient way to keep and use money wholesale, without the pressures of the international financial systems that are now putting some pressure on money laundering to Europe, Dubai, and North America.

Finally, whenever there is a manifestation of some of the effects of the cultural and societal devastation of Hausaland, that has transformed it from Africa's richest country to the poorest part of the world with the accompanying signs of the worst living conditions, the highest number of out of school children, and the highest rates of illiteracy found in just about any part of the world; those acts that should horrify everything that is decent and Godly, the standard dismissal and face-saving 'explanation' has become "oh well, that is not a Nigerian, it's a Nigerien," which is, of course, a comical and embarrassingly stupid attempt to try to explain away the sheer terrorism and criminality that has occurred in Hausaland over the last two centuries.

Let's take a look at some of these actions and events.

In July 2019, passengers on an Azman Air Flight, departing from Lagos to Port Harcourt were horrified as a man ran out of the bush beside the runway and climbed on the wing of the flight as it was about to taxi and take off. The man promptly threw his bag into the engine, and climbed on the wings of the air flight, waiting for the flight takeoff. The passengers on board were not impressed by this not-so-brilliant attempt at stowing away and raised an alarm thinking a bomb had been thrown into the engine and tried to get off the plane. The pilot also radioed the control tower for help. We will ignore the fact that it took almost 30 minutes for the airport security to reach the plane, and focus on some more enduring concerns. Quite expectedly and as has become the pattern, the Nigerian authorities issued a statement a few days later, that the action was committed by Nigerien, Usman Adamu and not a Nigerian. It sounded like they were almost giving themselves a pat on the back for the identity of the perpetrator. This story hits home on several points, as it has been explained here[222].

Well, the man was not insane; just completely ignorant about how you get on a plane and take a flight. A real tragedy on many levels and dimensions. This is because, in the 21st century, an adult male has not been exposed to civilization. You can however bet that he's certainly exposed to religion and can say his prayers. He can also be handed an AK47 and channelled to murder 'infidels' if and when necessary and so ordered by the Overlords of his country. There are millions, and possibly, tens of millions of Usman Adamus on both sides of Hausaland's Berlin-created Chinese wall/border which does not exist functionally, or based on the reality on ground. You thus have a large pool of young men, often devoid of any kind of education or access to

[222] (Alfa Shaban, 2019)

the benefits of modernity and civilization but are programmed, often with lies about a South that hates them, to fight or defend the cause of religion and kill infidels, if necessary.

This is the primary motivation for which invading clans have come from their homelands in Guinea, Mali, Mauritania and Guinea Bissau. The Fulbe overlords of Hausaland have destroyed previously existing and sustainable societies, and are now trying to extend the destruction they have wrought in Hausa Country to other countries and regions of Nigeria[223].

'Northern' Nigeria has over 2.9million internally displaced people which is caused by terrorism and poverty.

Adamu's case is one of many increasingly popular and continuous 'Blame it on Niger' syndrome that has completely overtaken Nigeria's ruling Fulbe clans, although 'Libyans' and 'Malians' (not 'Marlians', hopefully) hilariously sometimes get the blame as comical distinctions are attempted between local and foreign Fulbe, or town and bush Fulbe.

The 'Blame it on Niger' syndrome showed up again in February of 2021, during the Shasha market crisis in Ibadan[224].

223 (Africanews, 2020)
224 (Ajayi, 2021)

The crisis itself was unusual but not unexpected, given the repeated actions of Fulbe terrorists in Ibarapa in Northern Oyo State, although it should be pointed out that what might be termed as the trigger, the killing of Dr Fatai Aborode may have been an attempt to curtail the career of a rising political star by potential opponents, and not necessarily due to the Fulbe herders and terrorists who had previously threatened him and other commercial landowners in the area.

Sadly, given the ignorance of most of the Nigerian populace (and this is not suggesting that it is acceptable at any time for anyone to take the law into their hands), the killing of a trader by a Hausa trader or artisan in the market led to response and burning of stalls and killings and counter-killings by the indigenous market operators. Undoubtedly the actions of Fulbes in Ibarapa played a major role in this revenge action.

Sadly, this action was largely based on ignorance as most people outside of Hausa land do not know or care about the difference between the Hausa indigenes and their Fulbe overloads and do not realize and understand that Hausas are the most victimized and oppressed indigenous 'Nigerians'. So, anger and frustration over the terrorism of Herdsmen were vented on innocent Hausas, even though it should be noted, in what must be an extreme case of the Stockholm Syndrome, that Hausas have willing tools in the attack and killings of innocent people when directed by their overlords, on grievances flowing from alleged insults and disrespect of religious leaders, items, materials and institutions. There is however no basis, excuse or justification for attacking innocent people.

3000 people displaced and millions lost in the Shasha Market crisis.

What was most interesting though, was the official explanation given by the Nigerian authorities on the crisis which was that an artisan who was an indigene of the Niger Republic, struck and killed a local market operator, which led to the reactions and eventually the crisis in the market. Another case of 'Blame it on Niger'.

In June 2021, over 60 students and four staff members were kidnapped from a Federal Government College in Birnin-Yauri in Kebbi State, and expectedly the first reports about the identity of the perps, was ' *I don't think they are from Nigeria at all.* [225] Of course they are not. We know the standard narrative.

Niger is even blamed when camels end up colliding with cars, causing deaths on the highways[226].

Niger has also become a large recipient, custodian, and beneficiary of the fraud that is the Nigerian State, where a couple of Fulbe clans have morphed into a deep state that is first line of extraction of the resources from various parts of the country, especially the oil proceeds from the Niger-Delta., through various

---

[225] https://www.thecable.ng/i-dont-think-gunmen-who-attacked-my-school-are-nigerians-says-kebbi-student/amp

[226] https://dailypost.ng/2019/04/22/16-dead-foreigners-camels-cause-accidents-jigawa-roads/

acts of legally sanctioned or subversive corruption and transfer of resources.

As such, universities, railway lines, and even refineries are constructed, and established to continue the extraction and the milking of other Nigerian nationalities, which was exemplified by Nigerian officials signing contracts to guarantee the exportation of petroleum products from SINIDEP's Soraz refinery in Zinder, Niger[227] to a Nigerian market and citizenry that has been a victim of mind-blowing and unbelievable corruption from the NNPC over the decades.

NNPC and SINIDEP ratify a curious and inherently corrupt arrangement

[227] https://www.energyvoice.com/oilandgas/africa/pipelines-africa/280546/nigeria-niger-products-pipeline/

The deeply corrupt trend continues with the construction of a railway line with $1.8 billion added to the debts of already overburdened Nigerian states that will certainly only be paid for by people in the south and the middle belt to a Niger republic government that did not see the need for the project, and according to Transport Minister Rotimi Amaechi, they had to beg the Nigeriens to accept the project[228].

There have been many acts of direct corruption as well. The famous pension fund perp Abdulrasheed Maina was quietly reinstated by the Buhari administration, despite allegedly swiping a huge volume of funds. This information was eventually leaked to the press and caused a very public squabble between Winifred Oyo-Ita, the top civil servant in Nigeria's public service and Buhari's Chief of Staff, Abba Kyari.

Oyo-Ita was eventually fired from her role, despite a public make-up session with Abba Kyari and Maina conveniently crossed the border into the Niger Republic before he was reportedly brought back into the country. Another side note was the fact that a Federal legislator put up a house as bail security Maina's son, Faisal, who was also implicated on related money laundering charges. and the son conveniently skipped bail and left the country with the legislator losing the house in what seemed like a pre-agreed arrangement between all concerned parties as an easy way to get out of the situation, at least for the son.

228 (Olisah, 2021)

Abba Kyari. & Oyo Ita 'reconcile' following the leaking of a memo reinstating Abdulrasheed Maina in the Civil service despite unresolved criminal charges, despite the reconciliation, she was fired shortly afterwards,

In another scandalous and shocking but extremely underreported case of 'Blame it on Niger', 'Niger Republic' businessman Abubakar Hema was awarded weapons acquisition contracts of over $400 million and paid upfront. Although in reality most of these things are just money extraction arrangements. Hema did not deliver on the contracts and was declared wanted. This case again has been completely almost absent from mainstream Nigerian media, since deep Buhari Government insiders Mamman Daura, Abba Kyari and Isa Funtua have a longstanding media background, and have ensured that Buhari's government has been one that has a firm and almost complete control on traditional media. This is one reason why their limited control on social media irks them so much, and one reason why gestapo type powers are being concentrated in the hands of their younger (repentant under his own terms?) admirer of Bin-Laden, Isa Pantami.

In addition to the fact that there are credible Nigerian companies that could have been used for defense contracts, government-to-government arrangements could have been made, if the general objective was anything other than corruption and money laundering in the first place. So Fulbe overlords in the Niger Republic are great beneficiaries of the fraudulent setup of the Nigerian state. It's therefore not surprising that Muhammadu Buhari is an extremely popular figure in Niger and the candidates in that country reportedly use his posters at election rallies. That their benefactor's benevolence is nothing but the proceeds of the theft of other people's wealth and the destruction of their countries, is nothing but words from the unclean lips of infidels, and the fraudulent set up of the Nigerian State, and its so called constitution (a favorite word of those with violent invasion agendas, including Buhari and Pantami).

This affinity and popularity was confirmed as the leaders of some of the regions in the Niger Republic joined Buhari's campaign in Kano for the 2019 election. Maybe there are no laws on foreign tampering and interference for elections in Nigeria. Describing Nigerien interference as foreign in the same Hausa Country would be somewhat absurd anyway. Which makes it curious that foreign Fulbes are the ones with the loudest chants of 'One Nigeria' at all costs, without any restructuring, with subtle threats of another Civil war.

While all of this is happening on the Nigeria-Niger 'border', which is perfectly fine since Hausaland is one country, that need not be bound by the borders resulting from the land grabbing negotiations of the criminals in Berlin, the borders of Afeire Country in the southwest of Nigeria were permanently shut, and backed up with shoot on sight order, in what was probably a a genuine attempt at starvation, theft and an irrational transfer wealth from the south of Nigeria, which seems to be a personal obsession of Buhari's.

In normal climes, all the people responsible for the policy should be prosecuted and jailed. But Nigeria is not normal, and the fraud flows from the top and runs deep. That is why perps who murdered hundreds of their own colleagues and divided the country to depart from Nigeria, were never prosecuted, but were rather handed the 'two countries', and have continued to rule and pillage for almost six decades since July 1966.

Unsurprisingly, people claiming to be fighting for 'Northern' interests, quickly took the cue and and stopped commercial food laden trucks from crossing into Jebba, in the northern part of Afere country, after the Shasha crises, promising to starve the south, but forgot the 'sell it or you smell it' rule for perishable foods. Quite expectedly, they sheepishly ended their one sided fight and self imposed blockade within a week.

So unlike the overlords of the Nigerian State, we will not blame it on Niger. Hausa land is a naturally occurring country, and there should be no border splitting it into two. Same thing with the Afere Confederacy and the Borgu kingdom. As such, there should be no threats or war when stating the obvious regarding a nation created and held together for the benefit of foreign terrorists (colonial and violent propagators), who have continued to milk the system to benefit foreigners to the loss and outright pillaging of the indigenous Nigerians.

At the heart of this tragedy is the fact that Hausa land and indigenous Hausas have been the worst victims of the terrorism and violent propagation. Their Overlords came as peddlers of religion and cattle and were invited by the God-seeking rulers of Hausa land, to be priests in their courts. The usurping and takeover of all of Hausaland soon provided a platform for an almost complete destruction of the culture, society and economy, as the clans experimented with various forms of their extremism, which they could not practice in their homeland, and were meant to take to the banks of the Nile in the Sudan after 100 years.

Not surprisingly, the Fulbe conquest of Hausaland rapidly eroded the wealth and sustainability of the country, and Fuduye's son, collaborator and successor, Mohammed Bello, declared that it was okay to capture and trade the pre-existing indigenous Muslims as slaves, as they began to run out of non-Muslims to easily capture. The semi biographical accounts of the likes of Ayuba Suleiman Diallo and Samnuel Ajayi Crowther provides a window into those at the forefront and dominance of slave raiding, and Lugard's overthrow of the Fulbe leaders of Hausaland was predicated on their supposed refusal to end slave raiding.

The impact of Fulbe slave raiding in Northern Nigeria and the Trans-Saharan route was well known and justified by the perpetrators, but the impact of the invading Fulbes on the Slave trade in Southern Nigeria is often understated. While the accounts of Samuel Crowther provided a detailed window into this occurrence, the Fulbe destruction of (Old) Oyo has been said to be the key factor in the Trans-Atlantic slave trade, especially in the Afeire-Confederacy to which Oyo served as some sort of regulator. The impact on the people was said to be ***'catastrophic as they had largely avoided being made slaves up to that point but were now by far the majority of those captured and shipped to the Americas until the slave trade ended here in the 1850s CE'.*** [229]

No doubt, locals (including the well known and interestingly named Madam Tinubu) participated in the trans-Atlantic slave trade, but the major geo-political event that removed the previous restraint that had existed and triggered the explosion of the slave trade in south west of the current Nigerian state was the fall of Oyo.

[229] https://www.worldhistory.org/Oyo_Empire/

Ahmadu's Bello's reference of Tivs as his grandfather's slaves and repeated derogatory reference to Hausa's showed the forceful ending of slavery had little effect on the mindset of the invading rulers, who are able to achieve the same results (terror, financial remuneration, destruction of indigenous peoples and countries) through soul trading.

The Nigerien tries to catch a flight.

A traded soul like Usman Adamu who wanted to fly on the wings of the Azman aircraft, can be described as someone who is ignorant, uneducated, unable to function or live a productive life in the society or for himself, but has been programmed to follow a code (often under the guise of religion) and can be activated at the right time with an AK 47 for the cause of terrorism, or to kill 'infidels', and fulfill the self-serving objectives of a bunch of invading adventurist clans.

In no way can, or should this be blamed on Niger.

COMMENTARY 5 REFERENCES

Africanews. (2020, June 12). Retrieved June 17, 2021, from Africanews.com: https://www.africanews.com/2020/06/12/terrorist-attacks-increasing-across-northern-nigeria//

Ajayi, O. (2021, February 18). Retrieved June 17, 2021, from Vanguardngr.com: https://www.vanguardngr.com/2021/02/oyo-untold-story-of-shasha-market-clash/

Alfa Shaban, A. R. (2019, July 20). Retrieved June 17, 2021, from Africanews.com: https://www.africanews.com/2019/07/20/nigerian-man-attempts-physical-travel-on-wings-of-local-airline//

Bank, W. (2021). *Poverty & Equity Brief.* Washinton: World Bank. Retrieved June 17, 2021, from https://databank.worldbank.org/data/download/poverty/987B9C90-CB9F-4D93-AE8C-750588BF00QA/AM2020/Global_POVEQ_NGA.pdf

Olisah, C. (2021, March 3). Retrieved June 17, 2021, from nairametrics.com: https://nairametrics.com/2021/03/30/fg-explains-why-it-is-constructing-\rail-line-to-niger-republic-despite-criticisms/

SUBJUGATION AND THE NATIONAL IDENTITY

Russian writer Lev (Leo) Tolstoy is credited with saying that "All great literature is one of two stories; A man goes on a journey or a stranger comes to town." If this suggestion is true, then the history of the Berlin-mapped Nigerian State should make for some good reading, as two of the most central figures (Uthman Fuduye and Frederick Lugard) were strangers who rolled into town, with a desire for glory and the subjugation of the land and its peoples.

Both had good preparation. Fuduye's mentor, Jubril Umar of Agadez, had been expelled from Gobir for starting trouble. He had moved too soon, too quickly, and his initial strategy did not have mass appeal (no music or dancing for starters). Gobir was on the Western edge of Hausa Country, and was thus likely the first major city-state in Hausa Country that Fulbe pilgrims/migrants/tourists would encounter on their path to Sudan, or the Middle-East.

Significantly, many indigenous Nigerian countries and cultures suggest that it is a virtue to be kind to 'strangers' and travelers. An Afeire saying even likens the care with which strangers are to be treated to caring and raising Children (the most precious segment of society) with the saying that translates as 'Children are like strangers/travelers, it is our duty to show them the right way'. This duty and warmness towards 'strangers' also comes with a vulnerability, sadly.

Lugard had fine-tuned his glory-seeking skills as a sanctioned, but unofficial Muscle/Mercenary for English interests in Bugandaland, following his disgraceful exit from the British Army. The opportunity of raising a force in 'Nigeria' he could control outside of the official colonial channels, but with the presumed backing of the English Crown was too good to pass up.

Fuduye's more patient, wider net strategy eventually led to the capitulation of Gobir, and eventually with the reconnaissance and collaboration of cattle-bearing militia/herdsmen, the overthrow of the Sarkins of Hausa Country. The cattle bearing militia, posing as, or intermingling with herdsmen, soon crossed the Niger (or Kwara) River into Jebba, and soon got the consent of the King in the region, the Alaafin (Abiodun), to graze their cattle in the Country, as long as they did not go into the villages and towns, and did not damage crops in farms.

A few decades later, a future Alaafin soon lost his War Camp, Ilorin, and his Capital Oyo to the foreign Fulbe clans, but unlike the Hausa Sarkins, the Monarchy survived, as the Alaafin moved 130 Km South, to Ago or Atiba, where the local Chief, the Ashipa, quickly stepped aside, to become a chief in the Alaafin's council, and allow the creation of a new 'Oyo', where the stool of the Alaafin still sits as at 2021.

Not able or willing to maintain the norms and balance that made Hausa country the wealthiest in Africa before Fuduye rolled into town (which also made the vulnerable collection of city states desirable in the first place), Mohammed Bello, Fuduye's son and successor, settled for the 'low hanging fruit' of a slave raiding economy, that eventually had over 2 million people captured from Hausa country, and the surrounding areas, and progressively transformed the rich country into one of Africa's poorest. The Collapse of Oyo also exposed the people of that country to massive external and internal slave raiding and trading.

Fuduye believed or prophesied that his (clan's) adventure in Hausa Country would last for 100 years, and would be ended by another foreigner, so he gave instructions for onward movement to the banks of the Nile in the Sudan. Lugard, with a militia (Glover's Hausas) answerable to him, sidestepped his 'Mission to Borgu' to face Hausaland's slave raiding Fulbe clanleaders, somewhat fulfilling Fuduye's prediction, but then reinstalling the clanleaders. He eventually left for Hong Kong, and returned (for half the year, most of the time) to create the framework and amalgamation for what is by and large the current Nigerian State.

A third (self-identified) foreigner could be added to the mix. Even though Ahmadu Bello was born in Rabah, within the confines of Nigeria, he identified as having full Arabic origins in his brief autobiography, even if his Fulbe ethnic group are scientifically confirmed to be of Guinean and Senegambian identity.

Bello, having failed in his quest to be overall leader of the Fulbe ruling clans of Hausaland, (and escaping prison over a dispute about cattle at the hand of his clan leader and relative, thanks to the English Colonials), quickly realized that he could use the structure of the emerging Nigerian State to exert tremendous influence and control, even over the English colonials, despite their poor impression of his person. Bello and his clan members still maintained full control over what was now a poor and mostly ignorant Hausa Country, shed of its independent identity. The perfect place to easily get a million troops, if necessary, now that the empire had lost its crown jewel, India.

Completing our 'Tolstoy Quartet' of the stranger than fiction history is a man who may have rolled into town, as a child with his nomadic clans, or may have been born on the Nigerian-Niger border settlement of Dumurkul, as Mohammed Buhari Bafallaje claims. It is somewhat moot though, as Nigeria adopted everyone living within its borders, in 1960[230].

Maj. Gen Aguiyi-Ironsi, was accused by some as feeling the need to prove he was Igbo, due to suspicion and rumours of being Sierra Leonean. The 'Sierra Leonean father', was actually his older sister's husband, Thomas Johnson, but that did not seem to have much effect on the rumours, and some would say, the situation and decisions made.

Similarly, Muhammadu Buhari has had the need to 'prove' his Fulbeness to the ranking members of the ruling clans of Hausaland all his public life, though he seems to vacillate between being 'locked-in' as a Fulbe clan operative and executioner; and attempting to be a biological armistice of the centuries long war between the Fulbe ruling clans of Hausaland and his mother's Kanem-Bornu nation.

What cannot be disputed is that for Buhari, like the other foreigners in our quartet, Nigeria is subjugated territory, and the foreign Fuduyan Fulbe clans have a first right to its indigenous wealth and (human, capital, and and other natural) resources, irrespective of what the indigenous populace and countries think or say.

The 'Nigerian' Fuduyan clans also seem to have a need to show 'just how well they are doing in their conquered territory'

---

230 https://cadmus.eui.eu/bitstream/handle/1814/67855/RSCAS_2020_12.pdf?sequence=1

by exerting influence in their homelands, which is about 1000 kilometers from Nigeria. Being 'Nigerian' is a convenient afterthought.

That is why Buhari would completely ignore the youth protesting at the Lekki tollgate and host West Africa's Fulbe Presidents on the eve of the elections of their Guinean heartland. The loss of the Fulbe candidate in that election was followed by the bloodletting in Lekki. Could it be as a fallout of anger over the Guinean loss, or just coincidental timing?

Buhari, from his Dumurkul settlement origins and his time as administrator of the Northeast, must have been very familiar with the conflict between the Fulbe and Toubou, and decided the Chadian incursion provided the opportunity for him to use the might and resources of the peoples of Nigeria to propagate the Fulbe agenda to defeat their enemies.

In clear violation of his terms of engagement and reference and in clear and repeated defiance of his commander in Chief, he proceeded deep into Chadian territory with the clear intention to overthrow Chad's Toubou President, Hissene Habre.

Habre, Buhari, Shagari - May have dragged the Nigerian State into foreign ethnic disputes

Buhari has not been singled out for participating in the massive bloodletting of July 1966, especially since the killings of Brother-Officers and men evolved into a community offence by soldiers seeking (and getting, even if temporarily) a different country. It should however be noted that he has been coy about his involvement, or lack of it, especially since the Maryland-Ikeja Cantonment became the operational stronghold of the rebelling soldiers. It should also be pointed out that these July secessionists and killers of brother officers and men have mostly been the shapers and rulers of Nigeria in the five decades since 1975.

Nigeria has been very good to Buhari, well beyond his competence, and what he clearly deserves. Like the other Bello-Ribadu officers, who have severely punished (and plundered) the Nigerian state, qualification standards were waived to play catch up and dominate the Officer Corps of the military. He has been a provincial Governor, Petroleum Minister (then called Commissioner), Military Head of State (via a successful coup plot, so maybe Nigeria did not give him that) and Civilian President. There has been more than enough reason to charge him for treason and insurrection (although there may be the loophole of being a foreigner for these charges) and more, as he has threatened violence, death and chaos if he does not have his way.

Yet, Buhari seems to return stones and snakes for the bread and fish Nigeria gives to him. The fact he did not understand the most basic fiscal and operational issue in Nigeria's downstream Petroleum sector, (and by extension, one of the more significant public sector finance problems in the country), even as a former Petroleum Minister and Head of State, is a good indicator regarding why disastrous economic performances that have accompanied Buhari's rulership of the Nigerian State.[231]

Buhari's APC made restructuring the country one of the cornerstones of their electoral campaign in 2015. He was probably more concerned about actualizing his 'dogs and baboons soaked

[231] https://www.youtube.com/watch?v=YmG_dYY7YRA&t=29s

in blood' threat, to think through what restructuring meant. His party Chairman, nonetheless, commissioned a committee on Restructuring headed by competent administrator and Governor of Kaduna State, Nasir El-Rufai, a Fulbe known to be close to Buhari. The report, barely there on most of the key issues, was submitted to party Chairman, John Oyegun in 2018[232]. He was removed from office a few months later, and Buhari has stated that he does not know what restructuring means. Oyegun has kept calling for the Restructuring report to be implemented.[233]

Kaduna Governor El-Rufai submits the Party restructuring report to Oyegun

Meanwhile, Fulbe extremists and terrorists continue to stream into the country, with the generally held belief that Nigeria, beginning with Hausa Country, is conquered territory. Hausa Country has retrogressed to become the poorest and one

[232] https://www.premiumtimesng.com/news/headlines/256704-restructuring-apc-committee-recommends-state-police-resource-control-fiscal-federalism.html

[233] https://www.youtube.com/watch?v=OFpHGgcZE9E

of the least productive parts of the world (but a fertile greenhouse for extremism and terrorism) under the rule of the Fuduyan Fulbe Clans. This terrorism and misery is spreading, with a clearly identifiable source and purpose.

Sticking with El-Rufai, he tried to walk back a previous statement that could be interpreted as Fulbes killing security operatives by clarifying that: *"It is not a threat. It is not an incitement to violence and there is a context to it". ""The Objective was to remove them from the settlement because they are settlers and people that think that that land is their ancestral land, want to take the land. It is not because they had weapons. Now I know the nature of the nomadic Fulanis.*

"My great great grand parents used to be such people. If you kill any of them unlawfully, they organize to take revenge no matter how long it takes.

"Now for me, the danger is that if the Nigerian Army goes and wipes out a Fulani settlement, any person wearing a Nigerian Army uniform in 14 West African countries is at risk because these Fulanis may have relations in Mali, Ghana, Sierra Leone, Guinea, and word would go round that the Nigerian Army has created genocide against the Fulanis and anywhere you see a person wearing a Nigerian Army Uniform, is a target for retaliation and I warned the commander that this will happen." He did not listen and that is why I put it on the record so that when it starts happening, nobody would say he was not warned"

There is a whole lot to unpack from the statements above, and it could take a whole book to address, but it is almost certain that the welfare of the Nigerian soldier in other West African countries is the major factor or consideration here.

Also note that Governor El-Rufai is very well regarded by key power blocs in the Nigerian State, hence the decision by the ruling party to have him head its Restructuring Committee and try to appeal to Buhari's sentiments.

El-Rufai's statement/defense/justification touches on some of the more fundamental problems that have been left unresolved in the Nigerian state, including terrorism, sustainable societies, indigenous countries and identity.

His suggestion about Fulbes getting revenge for perceived slights and forced removals must be weighed against the facts and history to prevent chaos, lawlessness and anarchy, and to work towards achieving and maintaining a normal life within the context of sustainable and well-functioning societies.

Some Fulbe clans might be nomadic, some of the time, but they are not countryless. They are indigenous to at least six countries on the Western end of the African continent, about 1000 Kilometres away from any part of the current Nigerian State. These include Guinea, Guinea Bissau, Senegal, Mali, Gambia and Mauritania. The clannish set up of Fulbe society also means that Fulbes know what clans they belong to, and where the origins of those clans are, since Fulbe countries and regions are clearly identifiable, just as all Nigerians know the towns, areas and regions of their ethnic groups and sub-ethnicities. Orlu and Mbaise in the current Imo State and the Ijebu and Egba in the current Ogun State, for instance. Governor El-Rufai gives this away in his statements about 'relatives' in other countries.

It is likely that the Fulbe first encountered Hausaland seeking an alternative route to Arabia (Eastwards, then northwest around the Nile through Sudan, to Egypt or Ethiopia, then crossing the red sea to Yemen, or further north through Egypt) rather than go up north from their homeland and navigate through North Africa which has had recurring bouts of instability and could be fraught with dangers for travelers.

The rulers of the wealthy, stable, but often contending Hausa States were soon overthrown by their guests in Gobir. The basis for how the single incident led to the overthrow of all Hausa rulers is questionable, but Militia herdsmen, making threats all over Nigeria, have been known to claim that their right to Nigeria flows from their collaboration with the Fuduyan overthrow of Hausa rulers. Night time raids of villages and hamlets that kill the chiefs and leaders in indigenous communities, continue till this day (in 2021).

Mohammed Bello and his allies soon turned the Hausa States into full slave raiding enterprises. Over two million indigenous Nigerians, including Samuel Ajai (Crowther) were enslaved and sold, and many more displaced. Millions were killed. Almost all these people had no business or prior dealings with the Fulbe raiders.

You can imagine what would happen if these victims, or their loved ones and communities were to live by the Fulbe code suggested by Governor El-Rufai and revenge for each slave-raiding, killing and displacement.

Many of the Fulbe clanleaders came on a mission to kill, plunder and enslave people who had no prior dealings with them, and who had done them no wrong. Modibo Adama, for instance, clearly stated this intent with the Bata, and other indigenous groups of the Gongola region, with an intention to get to the Atlantic on the Eastern front. Yet, States, Universities and all sorts of monuments are named after these foreign invaders with a criminal intent to kill.

The Fulbe clans of Nigeria also curiously suggest that more Fulbes have migrated to Nigeria than remain in their indigenous countries of Guinea, Mali, Guinea Bissau, Senegal, Gambia in which they hold either the office of the President or Prime Minister (except for Guinea, where there may have been a military attempt to redress this bitter fact for Fulbes just before the elections[234]). The reality is likely to be less than half of the stated numbers, despite attempts to use examples of Fulbe leaders with 60 or more children (in 2021) as evidence of an increasing share of the population.The Fulbe deep State in Nigeria has a propensity to give a false impression of numerical strength. Many sites listed Nigeria's first Prime Minister as Fulbe, until called out on it. It is impossible that there will be more Fulbe in Nigeria than in their six or seven homeland nations combined, where they face no political, humanitarian, or environmental pressures to migrate.

It would not matter that much in a properly structured Federation, and population numbers would more accurately reflect reality, even if only for planning, and hopefully leaders would embrace less reckless behaviors as having 30, 60 or 80 children, or receive steep sanctions after reaching what society agrees is a reasonable limit, in a country struggling with extreme poverty, and a population explosion.

In addition to his pastime of slave raiding, Mohammed Bello also hosted would be violent propagators, Umar and Ahmadu Tall in Sokoto, lending credence to the suggestion that the Fuduyan Fulbe rulers of Hausaland viewed Hausa Country was a naive, low hanging fruit on the way to their initial Sudanese Nile destination.

Yet, somehow, the Fulbe Clan leader Rtd. Col Sa'ad Abubakar, a direct beneficiary of these activities, announces public holidays on behalf of the Nigerian state. The Nigerian Army, where he previously served carries the insignia of what was for indigenous Nigerians, a terrorizing foreign force. He has been honest to suggest that about 70% of the kidnappings in Nigeria are

[234] https://www.reuters.com/article/us-guinea-election-idUSKBN2711SD

by his clansmen. Yet all 'migrant and stagnant' Fulbes in Nigeria may not be more than 2% of the population. Extrapolating these numbers goes far beyond negative stereotypes.

No offense to Governor El-Rufai, but you do not want these nomads showing up in your town, village or forests, AK-47 bearing or not, especially since they have a network of 'relatives' that can show up to kill and plunder, when requested, and the track record does suggests killers will show up at the door shortly afterwards, especially for community or thought leaders.

As such, Muhammadu Buhari's insistence on initiating Ruga, Cattle Colonies and Grazing routes[235] reserves or Farming Estates for this foreign group in 2021 is nothing short of preposterous, even if he believes he has the Army, Police, State Security, and other collaborating enforcers to help him achieve what is clearly a sinister and brazen attempt to formally inject murderous foreign militia into every nook and cranny of the Nigerian State. Any farming Estates paid by the funds of the peoples of Nigeria must be under the control of the people. The indigeneous traditional rulers in the 24 non Sharia states can initiate agricultural trusts that control the Farming Estates in their area on behalf of their people for instance. Local indeginous youth can be trained in farming (include livestock farming) to professionally and sustainably manage the Estates.

[235] https://guardian.ng/opinion/buharis-obnoxious-open-grazing-routes/

Gumi Jr, and some of the 100,000 'bandits' sitting in front of a clearly identifiable UBEC (Basic Education) counterpart Niger State Funded project, which would have an address at both the State (SUBEB) and Federal levels. Yet, Military Officers and men had to die in reconnaissance flights looking for the bandits.

A key Fulbe leader in Nigeria, Abdullahi Ganduje, has asked all the herdsmen in other parts of the country to relocate to Kano[236], saying that there is enough space and facility to accommodate millions of herdsmen[237].

Rather than heed the call and see this donation of free land in Hausa country as an opportunity to become the beef and dairy hub of the nation, and maybe the sub-region, like Wisconsin or parts of California, the terrorist group (undeclared but what do you call a group that has boasted about killings) posing as herdsmen have declared that the foreign Fulbes own all the land in Nigeria.238 It is a claim that has been made many times in the past, and the basis for that claim is the supposed collaboration of the Fulbe herdsmen in the Fuduyan Fulbe takeover of the traditional, political and religious leadership of Hausa Country, and the supposed ceding of Nigeria to the Fuduyan Fulbes by Lugard.

If the Fulbe Herdsmen do not want to go to Kano as invited by Ganduje, a structured relocation program can be arranged (funded by indigenous Nigerian and if available, international donors) to relocate the herdsmen to their home countries, even if they go back and forth routinely. Guinea Bissau would be a choice destination to acquire all the needed land, since Embolo loves to visit Abuja so much, many times for very flimsy reasons that can hardly be justified.

There are reportedly up to 100,000[239] fully armed militia, with enough capacity to bring down military Aircraft, operating

---

236 https://punchng.com/relocate-to-kano-gov-ganduje-tells-herdsmen/

237 https://www.vanguardngr.com/2018/02/gov-ganduje-asks-fulani-herdsmen-relocate-kano/

238 https://www.msn.com/en-xl/africa/nigeria/all-the-lands-in-nigeria-belong-to-fulani-miyetti-allah-dares-governor-akeredolu/ar-BB1d1yhm

239 https://ng.opera.news/ng/en/society/ed1f5faa37fc07aa13a536aab1e04cf4

in the Northwest area of Nigeria and pushing Southwards. These terrorists only seem to be known by,and somewhat answerable to ranking Fulbe rulers (traditional, religious or societal) of Hausaland[240], with the Nigerian State, as represented by, or giving its backing to Gumi Jr. These group(s) should also be distinguished from Boko Haram/ISWAP (NorthEast) and the Fulbe Herdsmen (roving, Southern forest reserves, but with an open invitation to Kano).

These terrorists are demanding compensation for something, and many have suggested the terrorists were brought from Fulbe home countries (Mali has been mentioned) to back up the 'Dogs and Baboons soaked in blood' threat for the 2015 election, with the expectation that the APC Candidate would lose. Some have confessed to being armed by the Government.

The Nigerian State is very simple to understand. It has mostly been a collaboration between two sets of foreigners. It wasn't necessarily a collaboration of equals, but both parties were there for what they could get out of the countries, and not for what they could give. One group was from much farther away, about 5000 KM, and the other group has relatively closer homelands, about 1000KM from Nigeria. Lugard knocking off one Atahiru and replacing him with his similarly named brother, as head of the Fulbe overlords of Hausa Country in 1904 was the symbolic act of this collaboration between foreigners.

---

[240] https://thenationonlineng.net/fg-endorses-sheik-gumis-talks-with-bandits/

The makers and owners of the Nigerian State

One set of foreigners are gone, another set of foreigners believe it is time to subjugate the country, even if they have to destroy it to achieve that purpose. It is the classic case of the prostitute with the dead baby in Solomon's palace; the invaders will always use the, "I must have it or destroy it," approach. except that Fulbes are not countryless and have an open offer to graze in Kano, even if they insist their grazing has to be in Nigeria, rather than their well watered home country.

It does not take a genius to see the devastation that the Fulbe colonisation has wrought in Hausa country. The unwritten policy of death, slave raiding, destruction, poverty and illiteracy quickly turned Africa's richest country into its poorest.

The foreigner will work for his own interests and self preservation, without any consideration for what is optimal for the indigenous peoples. This is why:

- The Action Group was able to build more schools in three years, than the English Colonials built in Sixty;
- India's share of global GDP shrank from 17% to 4% under colonial rule.
- The Foreign Fulbe clan leaders will use extreme interpretations of religion to subjugate the populace of Hausa country in ways not practiced in their own countries, or by themselves, and absurdly claim immunity[241] as an excuse[242].
- The United States insists that its President must be a natural born Citizen. It is a country that loves to attract the right kind of foreigner, many of whom see it as the land of opportunity, but it places some political restrictions on these foreigners, no matter how exceptional they are in their fields.
- Moses was instructed, well before the end of the 'theocracy' that *you may indeed set a king over you whom the LORD your God will choose. One from among your brothers you shall set as king over you. You may not put a foreigner over you, who is not your brother.*

The hardly mentioned, but all encompassing and critical fault-line in Nigeria since the departure of the English colonials, is that the Fulbe overlords of Hausaland want to continue their

[241] https://gazettengr.com/we-only-arrest-poor-people-because-islam-frowns-at-public-criticism-of-leaders-kano-hisbah/

[242] ttps://dailypost.ng/2021/08/29/why-we-treat-rich-poor-differently-hisbah/

push downwards by any means possible, especially through the institution of the Nigerian State itself (Ruga, Grazing routes, NIN, Federal Character which invariably presents the tiny minority Fulbe overlords as the representatives in 12 or 13 Hausa States), while the Middle-belt and Southern parts of the country (in varying degrees and ways) say, 'No thank you, we see the devastating job of death, destruction, poverty and illiteracy that you've accomplished in Hausa Country.

This is quickly milked and spread as 'Hatred for the North," by the media savvy Fulbe elite, which has some credibility, but should be more appropriately viewed as a resistance to the spreading of the destruction of sustainable society to other 'countries' trapped in the Nigerian state.

This "Hatred for the North" narrative has also been an effective way to prevent the Hausa citizenry from questioning their Fulbe overlords, though Hisbah will have an excellent answer for those questions, should they arise, which might sound like 'Don't be sinful by criticizing, leave your leaders alone, overrun the hateful South' (Plateau, Benue and Niger State (Nupe Country) are on the frontlines as at 2021).

There is some element of victim blaming in there, but the thinking is that 'if the Hausas believe that they cannot find God for themselves and they need Foreigners (with no track record of building sustainable or desirable societies) to help them achieve this, while devastating their country, then that is fine, but the fallout of this decision should be limited to Hausa Country'. This is of course difficult in what is essentially a unitary Nation of different countries posing as a Federation.

A new political term, 'Hausa-Fulani' has actually forced its way into Nigerian official parlance.

MUFFLED MOUTHS, TRADING HANDS

Young Hausa men have been pouring out of Hausa Country to the South, since the Fuduyan takeover devastated Hausa country and economy. This is why it was easy to have the Lagos Constabulary (Glover's Hausa), dominated by ethnic Hausas.

Hausas are also quite enterprising and natural traders. 'Igbo trader' is a common Nigerian stereotype, but trading in Nigeria is somewhat segmented, and while Auto (Spare Parts) and Electronics Market is dominated by the Igbo trader, the food value chain is Hausa dominated.

Apart from controlling the sale of foodstuff in the markets, there are Hausa men operating corner shops in most streets of every city of Southern Nigeria, especially in the middle and lower income areas (Ikoyi and VI people don't want shops in front of their homes, to use Lagos as an example). Many of these men, that may number in the millions, came down South not speaking a word of English (making communication difficult), but have in the decades since established these shops that have become a primary or supplementary source of groceries and food items for most families.

A stall in front of a residence in Lagos (source - Al Jazeera)[243]

Most still retain their initial jobs as security guards or 'gatemen' and have lived in the 'BQ' of their employers for decades and now have their own families. The tendency to know a lot about events and people in the neighborhood has helped popularize the saying 'Wetin Musa no go see for gate'.

It is therefore not surprising that Hausa billionaires (and one time fierce competitors) Aliko Dangote and Samad Rabiu reportedly stand on family traditions over a Century of Kano to Lagos (Dantata) food and cash crops trade routes.

Besides neighborhood information, the 'Cornershop Gatemen' have also been an understated source of stabilization, in a Nigerian State that is becoming more fractured with all

243 https://www-fastly.aljazeera.com/programmes/my-nigeria/2015/08/nigeria-business-kiosk-150825085200283.html

kinds of pressures including the economic, population, social and identity issues.

Some of the current southward-bound set of young men are coming with more extreme and radical information than the age long 'Southerners hate the North' mantra. with shocking messages of Southerners stealing their wealth and being responsible for their poverty, and collaborating with Europeans to destroy 'them'. This is what makes many fear the outbreak of a long ethno-religious crisis at any time in Nigeria, as described by Bernard-Henry Levy, and several others.

Thankfully, the eyes believe more than the ears, and seeing many 'commoner' Musas as the mini social hubs in neighborhoods, instantly throws doubt to the narrative of Southerners that hate them. Many Southern youth, especially the educated ones, have given up on living in Nigeria anyway, and are actively plotting for a life of employment, constant electricity, potable water, no Police harassment, no Armed Robber, no Kidnappers and no Herdsmen harassment, potable water from a tap in Canada, Australia, New Zealand, Malaysia, Dubai, South Africa, or just about anywhere else where these necessities are guaranteed, since it is now almost impossible for Nigerians, apart from Doctors to emigrate to Western Europe and the United States.

Suspicions remain however, as many question how some of the new arrivals instantly buy motorcycles that cost about $1,000 to operate 'Okada' services. Some neighborhoods, especially those that hate to be classified as middle class, have called for the ban of these services in their areas[244].

There have also been recent (2020-21) incidents in Lekki and Ibadan (Shasha) that very quickly took an ethnic dimension. The

[244] https://pmnewsnigeria.com/2021/05/20/lekki-residents-express-worry-over-insecurity-call-for-okada-ban/

extorting activities of Park Unions (Road Transport Workers/ Employers) is an additional source of constant friction.

IDENTITY AND THE NATIONAL QUESTION

The identity question is a critical one for the Nigerian State. Hausa country is split into two on a map in what is effectively a borderless 'boundary' which doesn't and shouldn't exist[245] Maybe 15% of Afere country spills into the Republic of Benin, although you could be shot while trying to get to the other side of town to buy rice, in this case. A greater portion of Borgu is also in the Republic of Benin. The Eastern boundary is very interesting, especially below Kanuri country, as it is a global marvel and phenomenon, with over 10% of the world's languages, with about 100 of these languages each having less than 2,000 speakers[246].

This has often been used to fraudulently or ignorantly present Nigeria as a 200, 300 or 500 ethnic group mix that must be preserved as is, even though about a dozen ethnic groups living in contiguous and delineated countries make up over 95% of Nigeria's population.

There is no identity mess that needs a unitary government and identification system to resolve, outside of what the current crop of promoters are aggressively trying to hide or obscure.

Officially, the Chief promoter of the suddenly important NIN is Isa Pantami, the Communications Minister[247], supported

---

245 https://punchng.com/buhari-under-fire-says-only-god-can-police-nigeria-niger-border/

246 https://www.researchgate.net/publication/228865549_Moribund_languages_of_the_Nigeria-Cameroon_borderland/link/0fcfd5134d86a7a308000000/download

247 https://www.icirnigeria.org/again-pantami-extends-nin-sim-linkage-till-june-30/

by Umar Dambatta, the Head of the National Communications Commission, Aliyu Umar, Head of the Identity Commission, and Pantami's Personal Assistant who took over from him as Head of the Technology Development Agency[248], Kashifu Abdullahi; which gives the foreign Fulbes a 75% responsibility for Nigeria's national 'identity', in addition to the impressive 70% of the kidnappings. The identity number, might be higher considering the underbosses for the exercise are Buhari and very likely his older nephew and rumoured co-President (who lives in the State house) Mamman Daura.

A very impressive feat for the migrants who make up 1%, or 2% or the self reported 5% - 6% of Nigeria's population, who have now determined that 'Nigerians' cannot lead a regular life (get phones, passports, bank etc) in their own countries, unless they submit to the identity mechanism that these impressive immigrants, or migrant minority have determined.

The National Identity Card suddenly became important after a failed attempt to create Ruga camps, designed to take extremely large parcels of land in Nigeria's Federal states and plant foreign Fulbe herdsmen, already effectively a Fulbe killing militia. Some also reason this to be some sort of compensation for the Militia suspected to be imported for the post 2015 elections plans[249].

If this is true, then it will be similar to Sani Abacha escalating insecurity in the Niger-Delta March 1998 when Youth from the region were among those transported to Abuja to support the

---

[248] https://www.thisdaylive.com/index.php/2020/12/21/fg-extends-nin-sim-integration-deadline/

[249] https://www.thisdaylive.com/index.php/2021/03/07/is-buharis-government-behind-banditry-as-alleged-by-repentant-bandit/

(two) million man march[250] 'demanding' for an Abacha civilian (lifetime) Presidency. All the 'registered political parties' had already 'adopted' Abacha as sole candidate for their platforms, of course.

The Bello-Ribadu policy and the Officers it produced, can thus be likened to a poisoned chalice that has continuously brought a harvest of pain and corruption to the Nigerian States in the decades since.

As with most policies initiated by the Fulbe (no longer) deep state (with Buhari) in Nigeria, there was an attempt to sneak the policy through a fall guy[251], usually an indigenous Nigerian who couldn't be more different from the parties behind the policy (in terms of identity), with the victim this time being Buhari's Vice President[252]. Osinbanjo however found a voice to deny his consent to the policy supposedly passed through his office[253], even as his boss' aides continued to insist that he was their fall guy.[254]

The outcry over the Ruga policy was followed by the identity tinkering agency transfer, and after a pause of just under a year, Buhari renewed his obsession with snatching land from indegineous countries for foreign invaders by firing his Agricultural Minister, and quickly approving 109 integrated farming Estates in every senatorial zone of the country[255]. through a hardly known

---

250 Ebienfa, K. (2011). Militancy in the Niger Delta and the emergent categories. *Review of African Political Economy, 38*(130), 637-643.

251 https://www.thisdaylive.com/index.php/2019/06/29/miyetti-allah-drags-osinbajo-into-ruga-settlement-controversy/

252 https://dailypost.ng/2019/06/28/osinbajo-helping-herdsmen-create-fulani-ruga-settlements-nigeria-miyetti-allah/

253 https://guardian.ng/news/osinbajo-distances-self-from-ruga-settlement/

254 https://punchng.com/buharis-aide-contradicts-osinbajo-says-no-difference-between-ruga-livestock-plan/

255 https://thenationonlineng.net/buhari-approves-farm-estates-in-109-senatorial-districts/

Agricultural land Development Authority Agency. Buhari also started conversation about Grazing routes, despite widespread condemnation[256] and a consensus of legal advice on the illegality of such actions[257].

All of this, for herdsmen and clans whose homeland is over 1000 km from the Berlin created Nigerian State.

Grazing cows in a cornfield

[256] https://www.thisdaylive.com/index.php/2021/09/03/grazing-routes-herding-nigeria-into-needless-crisis/

[257] https://www.vanguardngr.com/2021/09/falana-hits-fg-no-grazing-route-anywhere-in-nigeria/

Complementing the land grab for foreign killer herdsmen[258] is the mechanism for the quick 'Nigerianization' of these foreigners that stroll across the 'borders', which is now perfected with the transfer of the 'Identity' agency to Isa Pantami.

The National Identity Management Commission (NIMC) has been around for about a decade and a half (it lists 2007 as its take off date), but for most people, it was just another agency that gets budgetary allocations, makes procurements, and rewards political patrons and power brokers.

The agency was just another evidence of the waste and multiplicity that plagues the Nigerian public sector, as many Nigerians have multiple forms of Government Identification. A Lagos resident can have a Federal Passport, drivers license, a LASRRA (State Residents) Card and Number, a Tax Card and Number, a Voter's Card and Number and Bank Verification Number (different from the 2 Bank Account (regular and NUBAN) Numbers. We'll disregard Staff ID cards, School cards, residents association cards, social club cards and other cards at this time.

There are also questions as to why a Federal agency should be coordinating this, since successive Governments have tried to deny the obvious, which is that the Nigerian State, is a unitary one trying to suppress its constituent components and countries, while its arbitrarily created branches masquerade as Federal units.

The Nigerian State has not addressed deep foundational and structural problems, so embarking on a National Identity card registration when 50,000 people can walk across the 'border' in certain parts of the country on any day and get 'Nigerianised' with the card is just a boneheaded case of putting the cart before the horse, without considering the likely sinister objectives.

---

[258] https://www.premiumtimesng.com/news/top-news/231380-herdsmen-attacking-nigerians-are-foreigners-minister.html

As such the sudden transfer of the NIMC to the Ministry of Communications in late 2020, and the rich language used to justify the move, was one of those things that make you go Hmmmmmm.

It should be stated that the official reasons given for the sudden importance of the NIN and phone SIM linkage have been quickly proven to the false[259], with the reverse seeming to be case, as phone networks were supposedly cut off in Zamfara State to fight insurgents, and not to track them[260]. The promoters now defaulting to new reasoning that it is the official means of identity authorized by (an act of) the National Assembly.

Most Nigerians had not heard of Isa Pantami, the Minister of Communications, a few months before then, but he hit the limelight following a 'power show' over an office building, or floor of a building, that was settled with the use of Armed security operatives,with former TV newscaster turned politician and Buhari's Diaspora Commissioner, Abike Dabiri, on the losing end of the tussle.

Nothing surprising about jockeying between the 'Ogas' and Madams, or about a Nigerian public official with multiple houses, which was reported shortly afterwards. The statement released afterwards seemed to deflect as much as clarify, but it certainly wasn't the Biafran Boys eyesore released by the Spokesman of the Code of Conduct Tribunal Chair[261].

If this was all that was newsworthy about Pantami, then there would be little to be concerned about, but the next few

[259] https://punchng.com/bandits-expose-sim-nin-linkage-failure-police-cant-track-gunmen/?utm_medium

[260] https://www.vanguardngr.com/2021/09/fg-shuts-down-telecom-activities-in-zamfara-over-insurgency/

[261] https://www.thecable.ng/extra-boy-was-rode-in-his-approached-ccts-gaffe-defending-assault-by-chairman

months revealed policies, information, and more importantly, indifference to information, that there might be no recovery from. Not for Pantami personally; as his strong backers will protect him; but for the Nigerian State itself.

Thanks to Pantami, anyone holding on to some dream or hope for some type of utopian manifest destiny or purpose for the Nigerian State, as currently structured and constituted, has been shocked and slapped right back to reality.

A voice recording of Pantami volunteering to lead what was essentially a killer squad to Langtang (the main town of the Taroks, who alongside the Biroms are the leading indigeneous ethnic group in Plateau State). This was related to the crises that have hit Jos and the surrounding areas over the last few decades, which have their roots in identity and the structure of the Nigerian State, in which self-sustaining countries have been forced into an unworkable unitary mess for the benefit of foreigners.

At no point in Pantami's emotional rally was there any talk of escalating his grievances to the police or any other official security agencies.

The Nigerian law only exists for the indigenes captured in it. The invading Fulbes have a militia or two to sort out those matters on their own terms, by wiping out villages for whatever reason anywhere in the country.

Widespread anger over the recording was largely ignored by Buhari, with a statement suggesting that the 'Minister was young, when the statements were made, and he has changed' being what the public was left to ponder on[262]. This was, however, unlikely to provide any comfort to the over 1 million people in

[262] https://www.vanguardngr.com/2021/04/controversial-terrorist-comment-im-now-mature-know-better-pantami-says/

IDP camps in the neighboring Benue, whose villages have been attacked by invading Fulbe Militia.

But more shocking allegations about Pantami soon came to light.

As an Assistant Lecturer and leader of the mosque in his university (significantly named after Nigeria's first Prime Minister, Abubakar Tafawa Balewa), he was said to have expressed anger over the evangelizing and tract distribution activities of other students[263], and the course of emotional sermons on the matter, declared a 'fatwa'[264].

The evangelising students were quickly expelled from the school, but apparently this was not enough for some, as the head of the ECWA fellowship (presumed to be the group of the tract distributors) in the school, Sunday Achi, a fourth year architecture student, was abducted and strangled.

It is not known if Pantami has ever been interrogated to give an account of what he knows about the death of a student that he should have been collaborating for communal interaction, and ensuring that peace and harmony on the campus of the University was maintained at all times. Bauchi, and Pantami's neighbouring claimed state of Gombe, are a part of the Nigerian State with a lot more diversity than most outsiders would imagine.

Somehow, there is no problem with an unneeded (certainly not before the structural issues of the Nation are resolved) identity scheme that seeks to control the lives of indigenous Nigerians, and Nigerianize invading terrorists is placed in the hands of Pantami and his former Personal Assistant.

---

263 https://punchng.com/pantami-was-imam-when-atbu-muslim-community-issued-fatwa-killed-my-son-in-a-mosque-prof/

264 https://www.icirnigeria.org/how-pantami-inciting-sermon-led-to-sunday-achis-death-former-atbu-students/

Dambatta's NCC has threatened to impose a N200,000 fine per line[265] for failure to link phones to 'their' NIN. These foreigners constantly default to the 'I have my way or destroy the baby/land/system/country/sustainable society' mode of Solomon's Palace, knowing that the indigenous peoples will likely surrender along the lines of Femi Adesina's 'Give up your land for ranching, and live' logic[266]

Pantami, Daura with Buhari

[265] https://www.thecable.ng/ncc-telcos-to-pay-n200000-fine-per-line-over-failure-to-verify-nin

[266] https://www.vanguardngr.com/2018/07/giving-land-ranching-better-death-presidency/

As a child, Abubakar Tafawa Balewa's grandmother told him nightly that when he got older, he was to 'kill the Fulani, and drive them out of their land'. Bitter and hateful words from a woman who may have been on the receiving end of Mohammed Bello's slave raiding and violent propagation. As an adult, Balewa expressed frustration and anger at the corruption and actions of just about everyone around him, but did nothing about it, believing that inclusion and accommodation was key to stability, and that Nigeria was not ready for opposition style politics (and presumably its checks and balances).

Maybe he was powerless and there was nothing that could have been done. Maybe we are all Abubakar Tafawa Balewa, in some sense.

Everyone blames the killings in Nigeria on Foreign Fulbe herdsmen. The question to ask is, just how the distinctions between local and foreign Fulbes are made, and how long it takes for a foreign Fulbe to become local?

Well, the answer will soon be right after they cross the border and are given the NIN number. The Malian 'Blood on Dogs and Baboon crew from 2015 must surely have now become local, with 100,000 militia, armed with anti-aircraft weapons waiting to be resettled in Ruga Camps and 'farming' settlements. Any mass killings on local populations by these invaders could be glossed over as just domestic disputes between herders and farmers.

There now seems to be an attempt to legitimize the takeover of the lands of indigenous peoples, which has been done through killings and terrorism over the last decades, but now by a government that is essentially made up of foreigners pursuing foreign interests, to the detriment of local indigenous peoples and countries. The foreign rulers of Nigeria are looking more and more like determined and unrepentant anarchists as at 2021.

It always seemed that 'subjugate or destroy' was a trade off and option that was given to the indigenous peoples, much like the 'split the baby or let me have it' scenario in Solomon's temple.

Right now, these options are fusing into one, with not much difference as the indigenous countries and peoples of Nigeria are faced with an invading group that seems determined to destroy the countries, even though it might believe it is here to subjugate it.

Muhammadu Buhari has stated that he doesn't understand what restructuring means, so he is not interested in restructuring the 'country'. He certainly isn't, and is doing a whole lot more than restructuring. He is on a determined course to destroy the existing ones.

Indigenous Nigerians have existed in their societies long before Pantami, Mamman Daura, Buhari, the suddenly important NIN, Ruga, Grazing routes, the English Colonials, the Fuduyan takeover in Gobir and the Nigerian State. Hopefully, these indigenous Nigerians, and their societies will emerge stronger from this difficult phase.

May God bless and protect the peoples and countries trapped in the current Nigerian State.

REFERENCES

ABOLADE, L. (2021, May 4). Again, Pantami extends NIN-SIM linkage till June 30. International Centre for Investigative Reporting. https://www.icirnigeria.org/again-pantami-extends-nin-sim-linkage-till-june-30/

Adepegba, A., Nnodim, O., Altin, M., & Isenyo, G. (2021, July 13). Bandits expose SIM-NIN linkage failure, police can't track gunmen. Punch Newspapers. https://punchng.com/bandits-expose-sim-nin-linkage-failure-police-cant-track-gunmen/

AfricaNews. (2020, June 12). Terrorist attacks increasing across northern Nigeria. Africanews. https://www.africanews.com/2020/06/12/terrorist-attacks-increasing-across-northern-nigeria/

Alfa Shaban, A. (2019, July 20). Nigerian man attempts to travel on wings of local airline. Africanews. https://www.africanews.com/2019/07/20/nigerian-man-attempts-physical-travel-on-wings-of-local-airline/

Ali, I. (2020, October 31). U.S. special forces rescue American held in Nigeria: Officials. Reuters. https://www.reuters.com/article/us-usa-nigeria-military-idUSKBN27G0JI

Buhari under fire, says only God can police Nigeria, Niger border. (2020, December 23). Punch Newspapers. https://punchng.com/buhari-under-fire-says-only-god-can-police-nigeria-niger-border/

Buhari's obnoxious open grazing routes. (2021, September 1). The Guardian Nigeria News - Nigeria and World News. https://editor.guardian.ng/opinion/buharis-obnoxious-open-grazing-routes/

Daka, T., Falaju, J., & Agbedo, O. (2019, June 29). Osinbajo distances self from Ruga settlement. The Guardian Nigeria News - Nigeria and World News. https://m.guardian.ng/news/osinbajo-distances-self-from-ruga-settlement/

Ebienfa, K. I. (2011). Militancy in the Niger Delta and the emergent categories. Review of African Political Economy, 38(130), 637–643. https://www.jstor.org/stable/23145870

Eniola Akinkuotu. (2019, July 2). Buhari's aide contradicts Osinbajo, says no difference between Ruga, livestock plan. Punch Newspapers. https://punchng.com/buharis-aide-contradicts-osinbajo-says-no-difference-between-ruga-livestock-plan/

Fg extends nin-sim integration deadline. (2020, December 21). THISDAYLIVE. https://www.thisdaylive.com/index.php/2020/12/21/fg-extends-nin-sim-integration-deadline/

Gen. Buhari (Rtd.) speaks on fuel subsidy removal. (n.d.). Retrieved 26 March 2022, from https://www.youtube.com/watch?v=YmG_dYY7YRA

Grazing routes: Herding nigeria into needless crisis. (2021, September 3). THISDAYLIVE. https://www.thisdaylive.com/index.php/2021/09/03/grazing-routes-herding-nigeria-into-needless-crisis/

Hines, C. (2021, January 15). Afghan Taliban leader bans polygamy for his commanders. SwordsToday.Ie. https://

swordstoday.ie/afghan-taliban-leader-bans-polygamy-for-hi s-commanders/

Https://www.vanguardngr.com/2021/07/announce-new-attah-igal a-yahaya-bello-told/. (2021, July 10). Vanguard Newspapers. https://www.vanguardngr.com/2021/07/announce-new-attah-igala-yahaya-bello-told/

Hunters Arrest Five-Man Kidnap Gang In Kogi. (2021, September 29). WonderTV. https://wondertvmedia.com/hunters-arrest-five-man-kidnap-gang-in-kogi/

Iroanusi, Q. (2021, September 21). Nigeria not interested in naming, shaming alleged terrorism sponsors—Presidency | Premium Times Nigeria. https://www.premiumtimesng.com/news/top-news/485917-nigeria-not-interested-in-naming-shaming-alleged-terrorism-sponsors-presidency.html

Is buhari's government behind banditry as alleged by repentant bandit? (2021, March 7). THISDAYLIVE. https://www.thisdaylive.com/index.php/2021/03/07/is-buharis-government-behind-banditry-as-alleged-by-repentant-bandit/

Jeyifo, B. (2000). On mazrui's 'black orientalism': A cautionary critique. The Black Scholar, 30(1), 19–22. https://www.jstor.org/stable/41069791

Jimoh, Y. (2021, August 23). The right person would be crowned as Attah Igala — Gov Bello. Tribune Online. https://tribuneonlineng.com/the-right-person-would-be-crowned-as-attah-igala-—-gov-bello/

John Oyegun: 'APC restructuring report should be implemented'. (n.d.). Retrieved 26 March 2022, from https://www.youtube.com/watch?v=OFpHGgcZE9E

Klasen, S. (2002). Low schooling for girls, slower growth for all? Cross-country evidence on the effect of gender inequality in education on economic development. The World Bank

Economic Review, 16(3), 345–373. https://doi.org/10.1093/wber/lhf004

Krumholz, W. (n.d.). Family breakdown and america's welfare system. Institute for Family Studies. Retrieved 26 March 2022, from https://ifstudies.org/blog/family-breakdown-and-americas-welfare-system

Mali approached Russian military company for help: Lavrov. (n.d.). Retrieved 26 March 2022, from https://www.aljazeera.com/news/2021/9/25/mali-approached-russian-private-companies-moscow-not-involved

Mazrui, A. A. (1991). Wole soyinka as a television critic: A parable of deception. Transition, 54, 165–177. https://www.jstor.org/stable/2934911

Miyetti allah drags osinbajo into ruga settlement controversy. (2019, June 29). THISDAYLIVE. https://www.thisdaylive.com/index.php/2019/06/29/miyetti-allah-drags-osinbajo-into-ruga-settlement-controversy/

More than 100,000 bandits in north-west alone, i need the fg to engage them—Ahmad gumi says—Opera news. (n.d.). Retrieved 26 March 2022, from https://ng.opera.news/ng/en/society/ed1f5faa37fc07aa13a536aab1e04cf4

Nasar, K. N. (2021, January 15). Taliban cracks down on 'costly' polygamy. BBC News. https://www.bbc.com/news/world-asia-55630097

Nwosu, A. (2021, August 29). 'Why we treat rich, poor differently'—Hisbah. Daily Post Nigeria. https://dailypost.ng/2021/08/29/why-we-treat-rich-poor-differently-hisbah/

Ochogwu, S. (2021, September 28). Benue: Herdsmen targeting us for elimination - Tiv traditional rulers raise alarm. Daily Post Nigeria. https://dailypost.ng/2021/09/28/benue-herdsmen-targeting-us-for-elimination-tiv-traditional-rulers-raise-alarm/

Ojoye, T. (2018, February 5). Relocate to kano, gov ganduje tells herdsmen. Punch Newspapers. https://punchng.com/relocate-to-kano-gov-ganduje-tells-herdsmen/

Olabimtan, B. (2021, March 31). EXTRA: 'Boy was rode in his approached' -- grammatical errors litter CCT's statement on assault saga. TheCable. https://www.thecable.ng/extra-boy-was-rode-in-his-approached-ccts-gaffe-defending-assault-by-chairman

Olajire, B. (2020, November 14). National common entrance cut off mark for unity schools 2021/2022 » servantboy. Servantboy. https://servantboy.com/national-common-entrance-cut-off-mark/

Olasupo, A. (2020, November 26). Nigeria retains position as the third most terrorised country in the world. The Guardian Nigeria News - Nigeria and World News. https://guardian.ng/news/nigeria-retains-position-as-the-third-most-terrorised-country-in-the-world/

Olatunji, H. (2021, September 6). NCC: Telcos to pay N200,000 fine per line over failure to verify NIN. TheCable. https://www.thecable.ng/ncc-telcos-to-pay-n200000-fine-per-line-over-failure-to-verify-nin

Olisah, C. (2021, March 30). FG explains why it is constructing rail line to Niger Republic despite criticisms. Nairametrics. https://nairametrics.com/2021/03/30/fg-explains-why-it-is-constructing-rail-line-to-niger-republic-despite-criticisms/

Olowoporoku, M. (n.d.). Lekki residents express worry over insecurity, call for Okada ban. Retrieved 26 March 2022, from https://pmnewsnigeria.com/2021/05/20/lekki-residents-express-worry-over-insecurity-call-for-okada-ban/

Olubajo, O. (2021, August 29). We only arrest poor people because Islam frowns at public criticism of leaders: Kano Hisbah. Peoples Gazette. https://gazettengr.com/we-only-arrest-poor-people-because-islam-frowns-at-public-criticism-of-leaders-kano-hisbah/

Onyekwelu, S. (2020, February 25). How cut off marks for Nigeria's Unity Schools may be destroying quality. Businessday NG. https://businessday.ng/education/article/how-cut-off-marks-for-nigerias-unity-schools-may-be-destroying-quality/

Opejobi, S. (2019, June 28). How osinbajo is helping herdsmen create fulani ruga settlements in nigeria—Miyetti allah. Daily Post Nigeria. https://dailypost.ng/2019/06/28/osinbajo-helping-herdsmen-create-fulani-ruga-settlements-nigeria-miyetti-allah/

Oyero, K. (2021, April 22). Pantami was imam when ATBU Muslim community issued fatwa, killed my son in a mosque – Prof. Punch Newspapers. https://punchng.com/pantami-was-imam-when-atbu-muslim-community-issued-fatwa-killed-my-son-in-a-mosque-prof/

Poverty and equity briefs. (n.d.). [Text/HTML]. World Bank. Retrieved 26 March 2022, from https://www.worldbank.org/en/topic/poverty/publication/poverty-and-equity-briefs

Rodriguez, J. (2010, August 30). Brick streets hold up better than asphalt • Aggregate Research International. Aggregate Research International. https://www.aggregateresearch.com/news/brick-streets-hold-up-better-than-asphalt/

Samb, S. (2020, October 16). Guinea military base attacked two days before presidential election. Reuters. https://www.reuters.com/article/us-guinea-election-idUSKBN2711SD

SHEHU, O. (2021, May 5). 'How Pantami inciting sermon led to Sunday Achi's death', Former ATBU students. International Centre for Investigative Reporting. https://www.icirnigeria.org/how-pantami-inciting-sermon-led-to-sunday-achis-death-former-atbu-students/

Shrimsley, R. (2021, April 16). The David Cameron scandal: Just how sleazy is British politics? Financial Times. https://www.ft.com/content/4362e62f-00fc-4cd9-a20b-9134bc2f0699

Sönnichsen, N. (2021, August 26). Oil production in Nigeria 1998-2020 . Statista. https://www.statista.com/statistics/265195/oil-production-in-nigeria-in-barrels-per-day/

Soyinka, W. (2000). The trouble with you, ali mazrui! - Response to ali's millennial 'conclusion'. West Africa Review, 1(2). https://www.africaknowledgeproject.org/index.php/war/article/view/422

Tukur, S. (2017, May 16). Herdsmen attacking nigerians are foreigners – minister | premium times nigeria. https://www.premiumtimesng.com/news/top-news/231380-herdsmen-attacking-nigerians-are-foreigners-minister.html

Ugbede, L. (2018, January 25). Restructuring: APC committee recommends state police, resource control, fiscal federalism | Premium Times Nigeria. https://www.premiumtimesng.com/news/headlines/256704-restructuring-apc-committee-recommends-state-police-resource-control-fiscal-federalism.html

Made in USA - North Chelmsford, MA
1334802_9781959024002
09.27.2022 1607